Practical Organic Synthesis

Practical Organic Synthesis: A Student's Guide

Reinhart Keese
University of Bern, Bern, Switzerland

Martin P. Brändle
ETH Zürich, Zürich, Switzerland

Trevor P. Toube
Queen Mary, University of London, London, UK

John Wiley & Sons, Ltd

Title of the original edition in German: Grundoperationen der präperativen organischen Chemie, 6th Edition, 2003, copyright © Professor Dr. Reinhart Keese, Department für Chemie und Biochemie, Universität Bern, Freiestrasse 3, CH-3012 Bern

Email (for orders and customer service enquiries): cs-books@wiley.co.uk
Visit our Home Page on www.wileyeurope.com or www.wiley.com

Reprinted with corrections May 2007

Library of Congress Cataloging-in-Publication Data
Keese, Reinhart.
   [Grundoperationen der präparativen organischen Chemie. English]
   Practical organic synthesis : a student's guide/Reinhart Keese, Martin P. Brändle, and Trevor P. Toube.
      p. cm.
   Includes bibliographical references and index.
   ISBN: 978-0-470-02965-7 (acid-free paper)
   ISBN: 978-0-470-02966-4 (pbk. : acid-free paper)
   1. Chemistry, Organic–Handbooks, manuals, etc. I. Brändle, Martin. P. Toube, T. P. (Trevor Philip), 1939- III. Title.
   QD257.7.K4313 2006
   547–dc22

                                                                            2006000913

British Library Cataloguing in Publication Data
A catalogue record for this book is available from the British Library

ISBN   978-0-470-02965-7 (Hardback) 978-0-470-02966-4 (Paperback)

Typeset in 9/12 pt Sabon by Thomson Press (India) Limited, New Delhi, India
Printed and bound in Great Britain by TJ International, Padstow, Cornwall
This book is printed on acid-free paper responsibly manufactured from sustainable forestry in which at least two trees are planted for each one used for paper production.

The present work has been carefully checked. However, the authors and publishers accept no responsibility for the accuracy of data, hints, advice, or bibliographical references, nor for any printing errors.

Contents

Preface

Basic laboratory techniques in organic chemistry are an essential element in the training of chemists, who must learn to employ their experimental skills in the preparative organic laboratory in order to undertake the safe, careful and successful synthesis of compounds.

This guide was originally developed in concept by R. Scheffold at the ETH Zürich as the basis for a successful course in practical organic chemistry. Further contributions by many teaching assistants, first at the ETH and later together with R. K. Müller, A. Pfaltz and R. K.'s students at the University of Berne, led to the publication of *Grundoperationen der präparativen organischen Chemie, eine Einführung,* now in its 6th edition, which has been used successfully as a teaching tool in the first laboratory course of preparative organic chemistry.

Many suggestions and comments from experts, colleagues and friends have augmented the information presented. Safety precautions and environmental considerations in the handling and disposal of reagents are nowadays significant concerns of the responsible experimentalist. The main chapters deal with practical procedures and guidelines for organisational aspects of preparing compounds. The chapter on searching the literature, illustrated by worked examples, is particularly useful in providing efficient access to chemical information; this material was developed by Martin Brändle, an information specialist at the ETH.

In addition to proving invaluable to the aspiring undergraduate, experience has shown that this practical guide is also useful to advanced students as a handbook or desktop manual in the laboratory: it contains many details about solvents, handling of air sensitive compounds, the synthesis and analysis of optically active compounds, and the disposal and deactivation of hazardous chemicals. Beyond the teaching aspects of this practical guide, we believe that our hints and suggestions may also provide a stimulus for chemists searching for solutions to practical problems in the preparative laboratory.

Mrs. H. Mischler-Brühwiler's cartoons and illustrations make this practical guide entertaining as well as instructive. Thanks are also due to P. Schär, who provided the pages from his laboratory notebook.

Practical Organic Chemistry contains a wealth of useful information, laid out in a clear, stimulating style, designed to appeal to novices and experts in this field. It is an essential guide to young students, a text to which they will return again and again throughout their careers.

Berne, February 2006   Reinhart Keese
                                                        Martin P. Brändle
                                                        Trevor P. Toube

Chapter 1

Accident Prevention and First Aid

1.1 SAFETY

Chemicals must always be handled with great care and attention. Many are potentially dangerous to health, possibly even poisonous and some are explosive. The majority of organic solvents are flammable, some have very low boiling points and may catch fire even on a hot surface at these low temperatures. Chapters 2 and 13 give details of the properties of some compounds.

It is essential that one is aware of possible hazards and prevents accidents.

1.1.1 Safety in the Laboratory

(a) *Laboratory Equipment*
    A typical, well-equipped laboratory has workbenches with cupboards beneath them, shelving, rotary evaporators and fume hoods with movable vertical shields at the front.
    There should be provision for leaving reactions running safely overnight.
    Separate waste bottles should be available for chlorinated and non-chlorinated solvents. Broken glass, chromatographic supports and drying agents should be collected separately (see Chapter 13).

(b) *Safety Apparatus*
    A laboratory should contain eyebaths, fire extinguishers as well as (often outside the actual laboratory) fire blankets, showers, bandages and fire alarms.

Practical Organic Synthesis: A Student's Guide   R. Keese, M.P. Brändle and T.P. Toube
© 2006 John Wiley & Sons, Ltd.

Before you start work make sure you know where the eyebaths, fire extinguishers, fire blankets, showers and bandages are located as well as the emergency exit routes.

Read the local fire rules and think about them, *e.g.* what happens when an alarm is set off? Always obey instructions from those in charge.

(c) *Personal Equipment*

   (i) *Laboratory Coat*

   Ideally one made of cotton, as synthetic materials can cause severe burns if there is a fire.

   (ii) *Safety Spectacles*

   Must be worn at all times. Contact lenses are a hazard, as solvents or other liquids can be drawn behind them by capillary action and then cannot be washed away quickly. Prescription spectacles are available for those who wear glasses.

   (iii) *Gloves*

   Rubber or latex gloves usually prevent skin contact with undesired substances.

   (iv) *Shoes*

   Wear closed, skid-proof shoes.

1.1.2 Some Rules About Conduct and Safety

(a) *Flames*

   • Bunsen burners: ideally, one should use a flame only when other heat sources are unavailable or inadequate. Bunsen burners should only be used in exceptional circumstances, and then in a fume hood. Even in such cases, it is essential that one makes sure that there are no flammable materials near the flame.

- Many fires are caused by inappropriate destruction of reactive compounds (see Chapter 13).

(b) *Smoking*
- Smoking is totally forbidden in the laboratory.

Forbidden

(c) *Glass Apparatus*
- Because glass is fragile, it must always be handled with care.
- Avoid increasing the internal pressure in vessels. Make sure that reaction or distillation apparatus is not a sealed system; use a drying tube or gas valve.
- Check that flasks do not have any cracks or flaws.
- Be particularly careful when opening ampoules; always cool in ice first.

(d) *Vacuum Work*
- Do not evacuate flat-bottomed flasks; they may implode.
- Use a safety cage when evacuating vacuum desiccators.
- Rotary evaporators can be made safer by enclosing the condenser in plastic netting.

(e) *Liquid N_2*
- Fill apparatus from the reservoir behind a safety shield.
- Wear safety glasses, gloves and solid shoes.
- Cover the open Dewar flask with a cloth when transporting it.
- Immerse cold traps in liquid N_2 slowly and under vacuum.

1.1.3 General Rules About Handling Chemicals

- A knowledge of the correct handling of chemicals and of potential hazards is an essential characteristic of a responsible chemist (see Chapters 2 and 15).
- Label all containers with a description of their contents. It is essential that contents are clearly identified.
- Always wear safety glasses and gloves when handling concentrated H_2SO_4. If acid gets onto the skin, wipe it off immediately with cotton wool or a cloth, rinse with plenty of water and then seek medical attention. To dilute concentrated H_2SO_4, add acid carefully to water (not *vice versa*). Sulfuric acid must not come into contact with chlorates, permanganates, concentrated ammonia or alkalis, or alkaline earth hydroxides, as they may react explosively.
- Contact with HF or its vapour is exceptionally dangerous. Burns are extremely painful. If you have to use HF, make sure you have read all the safety precautions first.
- Exceptional care is needed in handling carcinogenic substances:

  Do not use them except when unavoidable.

  Carcinogenic reagents and solvent can often be substituted by less hazardous substances.

  It is often possible to avoid the use of carcinogens by choosing alternative synthetic routes.

  Risks can be substantially reduced if one has the required skills and implements all the requisite safety precautions.

  Work in a fume hood, use disposable gloves, wear a dust mask and cover the work surface with aluminium foil.

  Handle carcinogenic material only in closed vessels.

  Destroy any excess reagents or reaction residues promptly.

- See Chapter 2: Environmentally Responsible Handling of Chemicals and Solvents.
- See Chapter 13: Disposal and Destruction of Dangerous Materials.

1.2 AVOIDING ACCIDENTS

1.2.1 First Aid

(a) *Eye Injuries*

If any chemical gets into the eyes, the eye should be washed at once with water at the nearest tap or eyebath. Washing should continue for at least 10 minutes. It is particularly efficacious and the injury becomes almost painless if the tap in the laboratory is fitted with a moveable mixing nozzle. The help of a second person in holding the head and holding open the eyelids to ensure thorough washing is desirable. This procedure is appropriate for accidents involving acids, strong bases (especially dangerous as they disintegrate the tissues and allow the contaminant to penetrate more deeply) and other chemicals.

Under no circumstances should acid splashed in the eye be washed out with bases (or *vice versa*), as this generally does more harm than good. After the eyes have been thoroughly rinsed with water for a sufficient period, professional medical attention should be sought. Contact lenses should not be worn in the laboratory as chemicals may be drawn under the lenses by capillary action and cannot easily be rinsed out.

(b) *Skin Burns*

Cool at once with running water. Larger burns may need to be covered with cloths soaked in cold water. These procedures also reduce the pain. Then seek professional medical attention. Do not cover burns with oily or greasy ointments. Do not puncture blisters.

(c) *Cuts*

Most cuts occur when handling glass. Many accidents can be prevented if glass is handled with a glass cloth or leather gloves, especially when being pushed through holes in stoppers, etc.

Take particular care to cool ampoules thoroughly in ice before opening them.

(i) *Treatment*

Small cuts:	Allow to bleed, disinfect, bandage.
Larger cuts:	If necessary stop bleeding, cover with a bandage (without disinfectant) and get medical attention.
Cut fingers:	Remove rings; unless they are obviously difficult, cuts on the fingers should be examined by a doctor.

(ii) *Bleeding*

Stop the bleeding by:

- raising the affected limb;
- if this is not sufficient, applying pressure at the appropriate pressure point (requires practice!) or applying pressure directly over the wound, using a large dressing.

Make sure bleeding has stopped.

Seek medical attention.

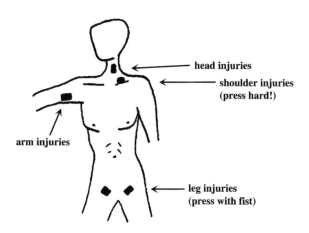

(d) *Poisoning*
Medical treatment is easier if the nature of the poison can be established (obtain a sample of the substance, gas, vomit, *etc.*).

General Procedure

• elimination of the poison;
• medical treatment.

(i) *Oral Poisoning*
Induce vomiting by giving warm salt water (three heaped teaspoons of NaCl per glass) to drink. Repeat until vomit is clear. Then seek medical aid.

Do not induce vomiting if unconscious or if solvent, acid or alkali has been swallowed.

If solvent has been swallowed, do not induce vomiting, but give $200\,cm^3$ of pure liquid paraffin to drink.

Then seek medical aid.

(ii) *Gas Poisoning*
Remove victim from the danger zone (wear compressed air breathing apparatus, if necessary), keep him/her quiet and transport to doctor on a horizontal stretcher. Painful coughing may be assuaged somewhat by inhalation of alcohol vapour (using a cotton wool pad soaked in ethanol).

(iii) *Percutaneous Poisoning*
Remove contaminated clothing immediately, wash the affected area thoroughly and then seek medical assistance.

In acute cases, the time factor is critical. Carry out the above first aid procedures and get medical attention immediately. Clothing must be washed and aired before re-use.

Knowledge of the toxicity of chemical substances is an essential part of the training of a responsible chemist.

Bibliography

[1] *Safety in Academic Chemistry Laboratories*. American Chemical Society, Vol. 1 and 2, 7th edition, 2003.
[2] R. S. Scott Stricoff and D. B. Walters. *Handbook of Laboratory Health and Safety*, John Wiley & Sons, 2004.

[3] R. J. Alaimo. *Handbook of Chemical Health and Safety*, American Chemical Society, 2001.

[4] *Safety Sense: A Laboratory Guide*, Cold Spring Harbor Laboratory, 2003.

[5] A. K. Furr. *CRC Handbook of Laboratory Safety*, CRC Press, 5th edition, 2000.

[6] W. C. Gottschall and D. B. Walters. *Laboratory Health and Safety Dictionary*, Wiley, 2001.

[7] *Prudent Practices in the Laboratory: Handling and Disposal of Chemicals*, American Chemical Society, 1995.

Links to information about safety in chemical laboratories

http://www.hse.gov.uk/chemicals is the official government source for the UK and is legally binding.

Online access to a large variety of information about safety (and many other aspects of chemistry and chemicals) is avaibable via PSIgate: http://www.psigate.ac.uk/ → chemistry → safety.

Information about occupational exposure to hazardous chemicals in laboratories is published by the Occupational Safety & Health Administration of the U.S. Department of Labor:

http://www.osha.gov/

Chapter 2

Environmentally Responsible Handling of Chemicals and Solvents

Practical Organic Synthesis: A Student's Guide   R. Keese, M.P. Brändle and T.P. Toube
© 2006 John Wiley & Sons, Ltd.

A wise head foresees problems!

Laboratory experiments should be planned to avoid danger as far as possible and to minimise the amount of refuse. Wherever possible, unreacted starting materials should be recovered and solvents recycled. Inappropriate handling of chemicals and solvents and careless behaviour must be avoided. One should also be aware of how to deal with any accidental mishaps which may occur.

It is essential that every chemist knows how to handle dangerous chemicals safely.

Before every experiment one should assess the possible dangers and health hazards, and consider what safety precautions are necessary. One should consider whether any of the substances involved are irritant, corrosive, flammable or explosive.

The following suggestions about appropriate experimental practice should prompt consideration of how one can protect oneself and others from danger, and awareness of the environmental effects of waste disposal.

The existing standards provide a minimum framework for safety precautions and procedures that should be observed in any laboratory. Obviously, local regulations supersede them in every case.

The following section merely indicates the immediate standards that need to be observed in a laboratory. It is intended to draw suitable attention to hazards and safety precautions.

2.1 DANGEROUS MATERIALS: POISONS, LAWS, REGULATIONS AND CLASSES

Poisons may be defined as substances which, by ingestion or contact, can endanger the health of people or animals by their chemical or physico-chemical effects even in relatively small amounts and which therefore need to be handled with particular care.

The framework for regulations for working with poisonous materials follows from this definition.

Poison Regulations describe the legal provisions covering poisons, prescribing in detail who may deal with them and what precautions have to be taken when handling them.

Poisons are subject to formal classification on the basis of their acute oral LD_{50} values in rats, but also taking account of data on toxicity, their effects on skin and mucous membranes, and absorption via the skin or by inhalation, as well as any available information on their behaviour in humans. A list of these poison classes is given in the following table.

2.2 EU POISON REGULATIONS

The European Union has guidelines and legal provisions aimed at harmonising the various national laws and regulations covering poisons and other dangerous materials, so that uniform grades can be established on toxicity, ecological effects and physico-chemical properties of substances.

Substances, including laboratory chemicals, are classified into three groups based on their toxicity and their potential danger.

EU *danger designations*	Swiss classifications
Very poisonous	1
Poisonous and corrosive	2
Health hazard; irritant	3 and 4

From 2005 new EU regulations came into force, incorporating the following R and S phrases and symbols (which are already widely used). A complete list is given in chapter 15.

Code	Hazard symbol	Danger descriptor	Safety precautions
T+ T		Very poisonous poisonous	Avoid any contact with the body as there may be serious health risks
C		Corrosive	Avoid contact with skin or eyes. Do not inhale vapour
Xn Xi		Health risk irritant	Avoid contact with skin or eyes. Do not inhale vapour
E		Explosive	Avoid sparks, heat, friction, shock or impact
F+ F		Highly flammable Flammable	Keep well away from flames and heat sources
O		Promotes burning oxidizing	
N		Environmental danger	

R Phrases (risk indicators)

Poisonous substances, including laboratory chemicals, are classified according to their potential hazards using the so-called R phrases. For example, R11 means 'highly flammable' [*e.g.* ether] and R34 means 'irritant' [*e.g.* phenol]. Combinations are also used: R23/24 means 'poisonous by inhalation or absorption through the skin'. These phrases assist hazard assessment.

S Phrases (safety indicators)
The particular risks indicated by the R phrases are supplemented by the safety advice given in the S phrases. For example, S8 means 'keep the container dry', S24 means 'avoid contact with the skin', and S36/37 means 'wear suitable gloves and protective clothing'.

The R and S phrases and/or the poison classes are now routinely indicated on the labels of laboratory chemicals and in catalogues of responsible suppliers. Substances for which such data are not available should be classified as R23/24 'poisonous by inhalation or absorption through the skin' as a precaution.

Conscientious experimentalists pay attention to these data; by this means, many possible accidents with chemicals can be avoided.

Examples

(1) **Dimethyl sulfate**

 T⁺

Handling information
on the label

R: 45-25-26-34
S: 53-26-27-45

R45 carcinogenic

R25 poisonous if swallowed

R26 highly poisonous by inhalation
R34 irritant

S53 avoid exposure; read precautions before use
S26 wash thoroughly with water if in contact with the eyes
S27 remove soiled clothing immediately
S45 get medical attention in case of accidents or if feeling unwell

(2) **Hexamethyl phosphortriamide (HMPA)**

 T⁺

Handling information
on the label

R: 26/27/28-45
S: 53-44

R26/27/28 very poisonous by inhalation, ingestion, or skin contact
R45 carcinogenic

S53 avoid exposure; read precautions before use
S44 may explode if heated in a sealed vessel

(3) Methyl iodide

 T+

Handling information
on the label

R: 21-23/25-40
S: 36/37-38-40

R21 danger to health by skin contact

R23/25 poisonous by inhalation

R40 may cause irreversible damage

S36/37 wear suitable gloves and protective clothing

S38 use breathing apparatus if is inadequate

S40 get medical attention if feeling unwell

2.3 CONCENTRATION LIMITS

The maximum limit of concentration of a gas, vapour or dust that, as far as current knowledge goes, can be experienced over a long period during a working day of 8–9 hours, or over a working week of up to 45 hours, without damage to health is designated as the OES (Occupational Exposure Standard) [TLV (Threshold Limit Value) in the US; MAK (Maximale Arbeitsplatz-Konzentration) in Germany, Switzerland and Austria]. It is measured in ml/m^3 (ppm). Concentration limits are also given by the workplace exposure limits (WELs).

The OES is not a definite boundary between safe and unsafe conditions. It also gives no indication of the dangers associated with short-term exposure to higher concentrations.

These figures are for exposure averaged over 8 h periods. In practice, the actual concentration of a gas or vapour or of dust may vary considerably over shorter periods.

SOME OES VALUES [ppm]

Acetaldehyde	50	Diethylaniline	10
Acrolein	0.1	Dimethyl sulfate	0.02
Aniline	2	Formic acid	5
p-Benzoquinone	0.1	Hexamethylene isocyanate	0.01
Boron trifluoride	1	Hydrogen chloride	5
Bromine	0.1	Hydrazine	0.1
Butyl acetate	5	Iodine	0.1
Chlorine	0.5	Maleic anhydride	0.2
Cyclohexene	300	Osmium tetroxide	0.0002
Diborane	0,1	Ozone	0.1
Dicyclopentadiene	0.5	Thionyl chloride	10

In order to avoid health hazards, the OES values should be considered in conjunction with information from toxicological studies concerning the effects of shorter exposure to substances at higher concentrations.

2.4 CHEMICAL CARCINOGENS

The transformation of normal cells into cancerous ones appears to be irreversible. A single molecular insult can lead to tumour formation. It also seems likely that carcinogenicity can result cumulatively from partial exposures. Because the latent interval between exposure to a carcinogenic substance and the appearance of a malignant growth can be very long, sometimes as much as 40 years in humans, the connection between the cause and the effect is not – as it is for acute toxicity – always manifested. A list of known and suspected carcinogens with access to the Material Safety data Sheets (MSDSs) is available from: http://ptcl.chem.ox.ac.uk/MSDS/carcinogens.html.

Some compounds which have been shown to have carcinogenic effects:

Alkylating agents

$(CH_3O)_2SO_2$ CH_2N_2 CH_3I

**Dimethylsulphate   †diazomethane   †methyliodide

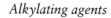

$ClCH_2—O—CH_2Cl$

**Ethylenimine   **β-propriolactone   †dichloromethylether

Solvents
†benzene   †hexamethylphosphoramidate
*1,2-dichloroethane   *carbon tetrachloride
*dichloromethane   *1,4-dioxane
*chloroform

Hydrazines and azo-compounds

$H_2N—NH_2$

†hydrazine   †N,N-dimethylhydrazine   **α-aminoazotoluene

Aromatic hydrocarbons

†benzene   †3,4-benzpyrene   **2,3,7,8-tetrachloro-
dibenzo-*p*-dioxin

Aromatic amines and nitro-compounds

†4-aminobiphenyl   †2-naphthylamine

†benzidine   **2-nitronaphthalene

Nitrosamines and nitrosamides

**N,N-dimethylnitrosamine   N-nitroso-N-methylurea

Natural products
These include aflatoxins, safrole, isosafrole, cycasine and pyrrolizidine alkaloids.

Inorganic compounds
†arsenic compounds
*asbestos (as dust)
†nickel and its compounds
*zinc chromate
**antimony trioxide
*chromium(VI) compounds

*beryllium and its compounds
**chromates
†nickel tetracarbonyl
**cadmium chloride (as breathable dust or
aerosol)

Others

**coal tar	**acetamide
**bitumen	†thiourea
*diesel motor emissions	†vinyl chloride
†beech sawdust	†oak sawdust

Carcinogenic substances and reagents, as well as those which are suspected of such activity, need to be handled responsibly and with care.

As it is not possible to ascertain with any certainty the safe concentrations for carcinogens, OES values are not published for them. However, as they are still used, legal limits are set, which provide the standards for preventative measures and monitoring. These limits do not define safe exposure values, as danger to health cannot be excluded even under such conditions.

Some examples:

- acrylonitrile   3 ppm
- benzene   5 ppm
- dimethyl sulfate   0.1–0.2 ppm
- ethylene oxide   1 ppm
- vinyl chloride   2 ppm

Biological exposure limits

The effects of substances can be measured in terms of atmospheric concentrations or by biological monitoring in the workplace. The biological exposure limit is the maximum permissible concentration of a substance or its metabolites at which exposure for 8 hours per day and up to 42 hours per week is, as far as is known, not likely to lead to any damage to health in the vast majority of cases.

Some examples:

- Cr(VI) compounds   20 µg Cr/l
- Ni (metal, sulfide, oxide, carbonate)   40 µg Ni/l
- acetone   80 mg/l
- butan-2-one   5 mg/l
- dichloromethane   1 mg/l
- dimethylformamide   15 mg/l
- methanol   30 mg/l
- tetrachloroethene   1 mg/l
- tetrahydrofuran   2 mg/l

---

*Suspect carcinogen on most recent results.
**Carcinogenic in animal tests.
†Known human carcinogen.
Absence of a cross means that no estimate of carcinogenicity is possible on present knowledge.

How to find information on dangerous materials and safety precautions?

For laboratory work, the most significant data are conveyed by the pictograms. The *R* and *S* phrases supplement these data with information about the specific substances.

This information can be found on the labels of all commercial laboratory chemicals and in the catalogues on responsible suppliers. Catalogues also have a 'General' section which contains an explanation of all these terms.

Bibliography

[1] R. J. Lewis. *Rapid Guide to Hazardous Chemicals in the Workplace*, Wiley-Inter-Science, New York, 2000.
[2] *Sittig's Handbook of Toxic and Hazardous Chemicals and Carcinogens*, P.P. Prohanish, Ed., Noyes Publications, Norwich, New York, 2002.
[3] R. S. Lewis and Van Nostrand Reinhold. *Hazardous Chemicals, Desk Reference*, 4th edition, New York, 1997.

Links to further important information and data

Material Safety Data Sheets (MSDS): An MSDS provides both workers and emergency personnel with the proper procedures for handling or working with a particular substance. The MSDS includes information such as physical data, reactivity, storage, disposal, protective equipment, spill/leak procedures as well as toxicity, health effects and first aid.
http://www.ilpi.com/msds/ref/ppe.html
http://www.ilpi.com/msds/ref.toxic.html

MSDS data can be found at: http://physchem.ox.ac.uk/MSDS/ → MSDS Information

Material Safety Data Sheets on CD are available from ALDRICH: Sigma-Aldrich MSDS (U.S., Canada, S. America), Sigma Aldrich MSDS Europe.

R(isk) phrases: "http://ptcl.chem.ox.ac.uk/MSDS/risk_phrases.html"
S(afety) phrases: http://ptcl.chem.ox.ac.uk/MSDS/safety_phrases.html

Workplace exposure limits (WELs): These values are determined for many chemicals on the basis of an 8 h exposure time. The list is published by the Health and Safety Commission in the UK and is legally binding.

http://ptcl.chem.ox.ac.uk/MSDS/wels.pdf

Chapter 3

Crystallisation

Practical Organic Synthesis: A Student's Guide   R. Keese, M.P. Brändle and T.P. Toube
© 2006 John Wiley & Sons, Ltd.

Crystallisation is one of the most effective purification techniques for solids (others include distillation and extraction). Crystalline compounds are generally more stable and easier to handle than solutions or oils, and can be effectively characterised and identified. Recrystallisation and fractional crystallisation, to constant melting point, as well as mixed melting point determinations, can serve as criteria of purity as well as aids to identification. Furthermore, good crystals are a *sine qua non* for an X-ray structure analysis.

Despite the development of many chromatographic methods, the preparation of crystalline derivatives and simple crystallisation still have a place in the laboratory, for example in the resolution of racemates via the preparation of diastereomeric compounds or salts [1].

Crystals can be obtained from the melt (supercooled liquid phase), the vapour phase (sublimation) or a supersaturated solution. It is crystallisation from supersaturated solution which is most commonly employed.

Before carrying out a crystallisation it is an advantage if one has some idea of how pure the material is and of the nature of the probable impurities. One can often use the normal methods for estimating purity (TLC, m.p., IR, NMR, *etc.*), although a closer study of the reaction (starting materials, side products, etc.) may need to be carried out. The same criteria of purity enable one to test the success of the crystallisation. Samples should always be recrystallised to constant mp. Products that stubbornly refuse to improve in purity may need sterner measures (chromatography, extraction, *etc.*).

What can crystallisation be used for?
• Isolation and efficient purification of solid products.
• Structure determination by X-ray or neutron diffraction.

Crystallisation can be carried out from the vapour phase (sublimation), solution or liquid phase (supercooled liquid). In physico-chemical terms it involves a change of phase. For controlling the procedure, one needs to know something about the solubility and vapour pressure of the substance. In terms of kinetics, there is a transition to the crystalline state from a supersaturated or supercooled medium.

The properties of some compounds may suggest particular procedures: many metalloorganic or aromatic compounds can be sublimed; crystallisation from solution is best for less thermally stable substances; formation of crystals from the melt requires significant long-term thermal stability, which is not always present. For these reasons, crystallisation from solution is the most common procedure. The following scheme is suitable for a simple crystallisation.

Plan of Work for a Crystallisation

Impure material
↓
(1)dissolve
↓
Solution
↓
(2)filter
↓
Filtrate
↓
(3)crystallise
↓
Crystals plus mother liquor
↓
(4)separate
↓
Damp crystals
↓
(5)dry
↓
Dry crystals
↓
(6)check purity
↓
Pure compound

(1) *Dissolution*

Preliminary tests for solubility can be carried out in ignition tubes using small amounts of the material and solvent, first in the cold and then heated. For crystallisation one should always use redistilled solvents.

Basic principle: Substances tend to be the most soluble in chemically similar solvents.

A list of suitable solvents and solvent mixtures for many classes of compounds can be found in [2]. Crystallisation is most efficient when the solubility of the compound at the maximum working temperature is 5–30% by weight.

It is a good idea to try different solvents to find the one which gives the most efficient crystallisation (*e.g.* using the series of increasing polarity as a guide). If the substance is already crystalline, do not use the whole sample for recrystallisation; save a few milligrams for seeding (and for TLC comparison, if appropriate). Dissolution takes time, and in some cases, the substance will have to be heated under

Class of substance		Efficient solvents
Hydrocarbons	Hydrophobic, lipophilic, non-polar	Pentane, hexane, petroleum ether
	↓	Toluene
Ethers	↓	Diethyl ether
Halohydrocarbons	↓	Dichloromethane
Tertiary amines	↓	Chloroform
Esters	↓	Ethyl (or methyl) acetate
Ketones and aldehydes	↓	Acetone
Phenols	↓	
Alcohols	↓	Ethanol
Carboxylic acids	↓	Methanol
Sulfonic acids	↓	Water
Organic salts	Hydrophilic, highly polar	

reflux in the appropriate solvent in order to obtain a sufficiently concentrated solution The purer the substance and the larger its crystals, the more slowly it will dissolve. Large crystals can be ground before adding solvent.

Weigh the crude material before crystallising so that the yield of purer material can be determined. The differences between the masses of crude and pure material should equal the mass of the material obtained by evaporation of the mother liquor.

(2) *Filtration*

Filtration serves to remove dust and insoluble impurities. It is often sufficient to filter the hot solution through a funnel fitted with a cotton wool plug. To prevent premature crystallisation excess solvent can be used, and the solution concentrated to the correct volume afterwards.

If the solution is strongly coloured by impurities, activated charcoal decolourisation is necessary. The material is first dissolved in a polar solvent (*e.g.* acetone, ethyl acetate, ethanol) and then treated with 2–4% of its weight of charcoal for 10 min with stirring (and if necessary heating). The solution is then filtered under suction through a pad of Celite (a filtration aid: a suspension of Celite is formed into a pad on the filter under suction; it holds back even fine particles of charcoal and avoids clogging the filter).

(3) *Crystallisation*

Supersaturated solutions needed for crystallisation can be prepared in the following ways:

- Slowly cooling a hot saturated solution to room temperature or below (using ice or refrigeration), which is suitable for the majority of substances that are more soluble in hot solvent than in cold.
- Slow evaporation of solvent (or the better solvent if using a mixture) in a fume hood or using a cold trap.
- Slowly adding a miscible poor solvent to a solution until it just starts to go cloudy (mixed solvent technique). Typical mixed solvents are dichloromethane–hexane, chloroform–hexane, ether–hexane, ether–acetone, acetone–water, and methanol–water. If possible choose a system in which the better solvent is the one with the lower boiling point.

Crystallisation often refuses to occur spontaneously even when the solution is supersaturated, and must be induced by the formation of crystal nuclei. This is often difficult and may occur only with highly supersaturated solutions, especially if the compound is very impure or the solvent system unsuitable, or not at all!

The following techniques may help:

- addition of large or small seed crystals that have been washed with a solvent in which they are barely soluble;
- scratching the side of the vessel with a glass rod;
- repeated immersion in an ultrasonic bath;
- sudden cooling with, *e.g.* solid CO_2; and
- addition of an unrelated, insoluble substance.

Rules of thumb:

Maximum formation of nuclei occurs *ca.* 100 °C below the mp.

Rapidity of crystallisation is greatest *ca.* 50 °C below the mp.

It is relatively ineffective to leave a viscous mixture in the refrigerator for a long period; temperature cycling is more likely to succeed. This is because the highest rate of nucleation is likely to occur at a temperature well below that for crystallisation to proceed most rapidly. Thus, sudden cooling for a short period may produce nuclei that will grow more efficiently on warming. If the substance comes down as an oil, one can try crystallisation from more dilute solution or with less rapid cooling or at a temperature 50–100 °C below the expected melting point. Important aspects of crystallisation from solutions are discussed in [3].

Fractional Crystallisation

This is a technique for separating two substances by repeated crystallisation (*e.g.* separation of racemates via their crystalline diastereoisomeric derivatives).

A schematic representation of the process is shown below (K ≡ crystals, ML = mother liquor).

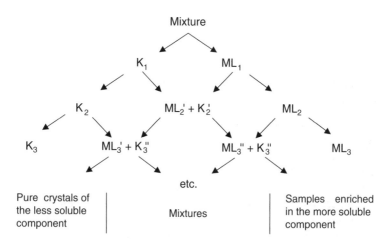

(4) *Separation of Phases*

The mother liquor is removed from the crystals by filtration, preferably through a glass sinter, or under an inert gas (nitrogen, argon) in a special filtration apparatus. For small samples it may be possible to draw off the mother liquor using a finely drawn pipette.

The crystals should then be washed quickly on the filter with cold solvent of the same composition as the mother liquor; if this is not done in the case of mixed solvents, the single solvent used will either dissolve the crystals or else precipitate impurities from the traces of mother liquor still clinging to the crystal surfaces!

(5) *Drying*

Crystals should be dried to constant weight in a desiccator or other suitable vessel, usually under reduced pressure (water pump or high vacuum with cold trap). Thermally stable compounds can be warmed to *ca.* 50 °C below their mp as long as their volatility is not too great (*i.e.* excessive drying under vacuum can lead to loss of material by sublimation). Drying agents should only be used in desiccators when crystals from aqueous solvents are being dried.

When drying crystals, beware of decomposing solvates. Figures show that the solvents most likely to form solvates are water, dichloromethane, benzene and methanol.

(6) *Tests for Purity*

The efficacy of a recrystallisation may be judged using one or more of the following criteria.

Melting Point (mp)

The mp is the most common test for purity after recrystallisation. Even small amounts of impurity may depress the mp appreciably. In general, one recrystallises to constant mp. A further increase in purity may be obtained by sequential recrystallisation – it may pay to crystallise from an alternative solvent system. Material should be carefully dried and finely pulverised for mp determinations. An estimate of the mp range should first be made, heating the sample comparatively rapidly. A second sample is then heated more slowly (*ca.* 2 °C per min) from about 20 °C below the estimated value to obtain the exact mp. Melting points are usually determined in capillary tubes (mp tubes) open to the air. In special circumstances sealed evacuated capillaries may be necessary. A heated stage microscope (Kofler block) is sometimes used especially when the mp of a very small sample is required.

The result may be quoted thus: 'mp 109–110 °C (uncorrected, open capillary), lit.[ref.] mp 110 °C'.

Mixed mp

A mixed mp serves to identify a substance. An intimate 1:1 mixture of the unknown compound and of a pure sample of the reference compound is prepared. Three mp tubes are used containing the unknown (A), the reference (B) and the mixture (M), respectively, and they are heated at the same time in the same apparatus.

  (i) If A, B and M melt essentially simultaneously, the unknown is identical to the reference compound and equally pure. (Strictly, the two compounds should have the same crystalline form and unit cell.)

(ii) If A melts first, then M followed by B, the unknown may be an impure sample of the reference compounds.

(iii) If M has the lowest mp (mp depression), A and B are different compounds (even if they fortuitously have identical mp).

The advent of effective spectroscopic methods for the identification of organic compounds has decreased the traditional use of crystalline derivatives. However, crystalline derivatives are still a useful aid for identification. A list of such derivatives for the various classes of compounds can be found in, e.g. [8].

Crystals are also a prerequisite for X-ray or neutron diffraction methods for structure determination. These usually require crystals with linear dimensions of about 0.3 mm. Preparation of such crystals follows the procedures outlined above, but the conditions for nuclei formation and crystal growth need to be controlled more carefully. In addition, these crystals should not be dried in a vacuum, as escaping solvent may distort the ordered lattice structure [3]–[8].

Bibliography

[1] J. Jacques. A. Collet and S. H. Wilen. *Enantiomers, Racemates and Resolutions*, Krieger Publ. Co., Florida, 1991.

[2] C. Reichardt. *Solvents and Solvent Effects in Organic Chemistry*, 3rd edition, Wiley-VCH, 2002.

[3] J. Hulliger. *Chemistry and crystal growth, Angew. Chem. Int. Ed. Engl.*, 1994, *33*, 143–162.

[4] Techniques of Organic Chemistry, 3rd edition, R. S. Tipson, in A. Weissberger, Part 1: Separation and Purification, 1966.

[5] J. W. Mullin. *Crystallisation*, Butterworth-Heinemann, 4th edition, 2001.

[6] K. Byrappa and T. Ohachi. *Crystal Growth Technology*, Noyes Publications, 2003.

[7] R. J. Davey and J. Garside. *From Molecules to Crystallisers*, Oxford University Press, Oxford, 2001.

[8] C. K. F. Hermann and R. L. Shriner. *The Systematic Identification of Organic Compounds*, 8th edition, Wiley-VCH, 2003.

Chapter 4

Distillation

Distillation is one of the chief techniques for separating mixtures of liquids. Separation is based on the difference in vapour pressure of the various components of the mixture. The mixture is vaporised by heating and the vapour is then condensed, and in the process the vapour (and therefore the condensate) becomes enriched in more volatile components.

Practical Organic Synthesis: A Student's Guide   R. Keese, M.P. Brändle and T.P. Toube
© 2006 John Wiley & Sons, Ltd.

4.1 SOME THEORETICAL CONSIDERATIONS

4.1.1 The Clausius–Clapeyron Relationship

The vapour pressure (P) of a liquid rises rapidly with temperature (T). If the molar heat of vaporisation is L_v, the Clausius-Clapeyron relationship gives

$$\frac{\mathrm{d}\ln P}{\mathrm{d}T} = \frac{L_v}{RT^2} \tag{4.1}$$

A graphical representation of an integrated form of (4.1) shows a straight line the slope of which is determined by the molar heat of vaporisation (which does not differ greatly between chemically similar materials). If one knows the boiling point (bp) of a liquid at a particular pressure, the bp at any other pressure can be estimated (L_v is taken to be temperature-independent in this treatment).

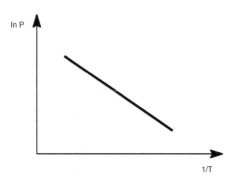

Figure 4.1: Graph showing the relationship between boiling point and pressure

Rules of thumb

• Halving the pressure reduces the bp by about 15 °C.
• Water pump vacuum (10–15 Torr) usually reduces the bp by about 100 °C.
• Vacuum pump pressure (*ca.* 0.1 Torr) causes a further reduction of about 60 °C.

4.1.2 Raoult's Law

(a) *Simple Distillation*
The enrichment of the more volatile component in the vapour of a binary mixture after a single evaporation is given by Raoult's law as:

$$\frac{y}{(1-y)} = \frac{P_a}{P_b}\left\{\frac{x}{(1-x)}\right\}$$ (4.2)

where y = proportion of more volatile component in vapour
  x = proportion of more volatile component in container
  $P_a\, P_b$ = vapour pressures of pure components a and b.

PRACTICAL EXAMPLE

With a 1:1 binary mixture (i.e. $x = 0.5$) having a bp difference of 60 °C (giving $P_a/P_b \approx 10$), solution of equation (4.2) gives $y = 0.9$ (i.e. in the vapour there is a ratio of 9:1 in favour of the more volatile component, a).

IMPORTANT RULE OF THUMB

For simple distillation, tantamount to a single vaporisation process, to be really effective the components should differ in their boiling points by at least 80 °C.

In practice this means that simple distillation should only be used to separate an already fairly pure volatile substance from high-boiling impurities (*e.g.* solvent from inorganic impurities or drying agents, reaction products from polymeric by-products, *etc.*). Equally, it can be used to remove all solvent from certain liquid reaction products.

(b) *Fractional Distillation (Rectification)*
If the bp difference is too small for simple distillation to be effective, it is necessary to resort to repeated distillations. In practice one employs a **fractionating column** in which the vapour and condensed phases move in opposite directions. The efficiency

of such columns is expressed in **theoretical plates**. A theoretical plate is defined as the column unit having the same effective separation as a simple distillation (and often expressed in centimetres of column height).

For an n-fold repetition of the vaporisation–condensation process, the enrichment of the more volatile component is given by

$$\frac{y}{(1-y)} = \left(\frac{P_a}{P_b}\right)^n \left(\frac{x}{(1-x)}\right)$$

(4.3)

where x, y, P have the same meanings as in equation (4.2).

PRACTICAL EXAMPLE

With a 1:1 binary mixture (*i.e.* $x = 0.5$) having a b.p. difference of 30 °C (i.e. $P_a/P_b \approx 3$), solution of equation (4.3) to give $y = 0.95$ requires $n \approx 3$, i.e. one would need a column of at least three theoretical plates to get the more volatile component at least 95% pure.

4.2 DISTILLATION IN PRACTICE

4.2.1 Choice of Temperature and Pressure

An unknown thermally stable mixture should first be simply distilled at atmospheric pressure. Heating should not be continued above about 180 °C even if material that will not distil remains in the flask. The distillation flask is cooled, the collected fractions are removed and the distillation is continued under reduced pressure. Higher boiling material should be distilled using a vacuum pump at *ca.* 0.1 Torr. Heat-sensitive, unstable compounds should, of course, be distilled only under reduced pressure.

4.2.2 Apparatus

The choice of apparatus depends fundamentally on the nature of the problem, as shown by b.p. differences, types of impurity, amount of material *etc.*, for some suggestions, see the illustrations.

Apparatus for simple distillations

Round-bottomed flask
with K adaptor and
coil condenser

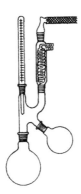

Suitable for
distillations of solvents
at normal pressure

Round-bottomed flask
with Claisen head and
receiver bend

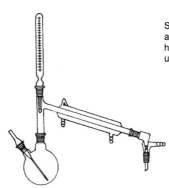

Suitable for separation of
amounts (>10 g) from a
high-boiling residue. May be
used for vacuum distillation

Bulb (short path) distillation
apparatus (kugelrohr)

Suitable for distillation
of small amounts
(ca. 100 mg – 10 g). May be
used for vacuum distillation

Apparatus for fractionation:

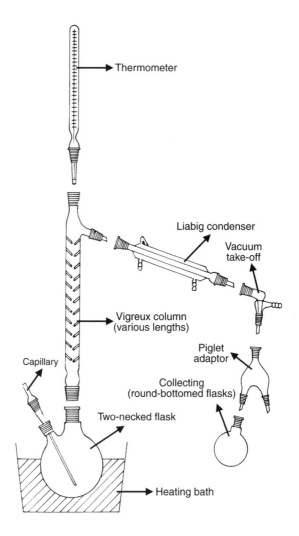

The figure shows the standard apparatus for fractional distillation.

More effective separation requires the use of other columns (spinning band, packed, etc.) and a head allowing the reflux ratio to be adjusted.

4.2.3 Heating

A waterbath can be heated to 100°C, liquid paraffin to 135°C and silicone oils to 180°C (some will remain stable for a short time up to 250°C), while metal baths can be used for higher temperatures. Heating mantles are only suitable for distillation of solvents or for heating thermally stable compounds under reflux; they are not easy to regulate and transfer heat irregularly and are thus dangerous.

Avoid bumping
Normal pressure – anti-bumping chips, boiling sticks, platinum tetrahedra, magnetic stirrers.
Vacuum   – capillaries, magnetic stirrers.

4.2.4 Flask Size

It should be about 1.5 times the sample volume.

4.2.5 Columns

Match the column size to the quantity of material; large columns lead to excessive wastage. The efficiency of columns depends on various factors, which are given below.

Column	Diameter (mm)	Load (ml h^{-1})	Theoretical plate height (cm)
Empty column	6	115	15
Vigreux	12	194	7.7
Packed column	24	100–800	6
Spinning band	5	50–100	2.5

- column type
- column loading (it must not 'choke')
- constancy of pressure
- insulation (cotton wool, silvered vacuum jacket)
- reflux ratio (very important, but only adjustable if special apparatus is used)

4.2.6 Thermometer

The top of the mercury reservoir should be opposite the side arm leading to the condenser.

4.2.7 Cooling

Pass cold water through the condenser. If the distillate solidifies in the condenser, use warm water (water bath and circulating pump) or change to an air condenser.

4.2.8 Receiver Adaptor

If it is necessary to change receivers during distillation without breaking the vacuum, a special adaptor should be used (*e.g.* a 'spider' or 'piglet,' or a Perkin triangle).

4.2.9 Receiver

Weighed round-bottomed flasks (essential for measuring yields) should be used.

4.2.10 Cold Traps

They are essential for high-vacuum work (to protect the pump from contamination by volatile substances).

N. B. Even at $-78\,°C$ some solvents may be volatile under high vacuum; predrying at room temperature using a water pump usually removes these materials.

4.2.11 Pumps

It is best when using water pumps to fit them with a vacustat, so that water consumption can be minimised and contamination of the waste water by solvent (especially dichloromethane) avoided. Dry-sealed membrane pumps, which permit the condensation of volatile solvents on the compression side, are preferable.

4.2.12 Cold Bath

A Dewar flask of solid CO_2 in propan-2-ol is effective.

4.2.13 Manometer

It is used to measure pressure, but not needed for distillation at atmospheric pressure (in which case a drying tube or N_2 reservoir may be attached to the apparatus to protect the products).

4.2.14 Distillation Record

A record should be kept for every distillation showing the weights of all fractions.

Example

Material to be distilled 20.2 g.

Fraction	Bath temperature (°C)	Vapour temperature (°C)	Pressure (Torr)	Weight (g)	n_D	Remarks
1	80	55	760	8.3	1.444	
2	60–75	22–55	18	1.1	1.416	Mixture
3	75–80	59	18	6.6	1.409	

4.2.15 Azeotropes

There are certain binary or ternary mixtures of liquids which are not separable by distillation even though the boiling points of the components are sufficiently far apart. Such mixtures give a vapour of *constant* composition at the so-called azeotropic boiling point (usually below the bp of any of the components – a 'minimum azeotrope'). This azeotropic mixture (or constant boiling mixture) will continue to be distilled as long as the quantities of the components in the flask remain sufficient. On condensation the azeotrope may separate into two liquid phases.

Components			Azeotrope		
			Composition (%)		
	bp (°C)	bp (°C)			
			Azeotrope	Upper phase	Lower phase
Water	100	56.3	3.0	99.2	0.2
Chloroform	61.2		97.0	0.8	19.2
Water	100	85	20.5	0.05	99.94
Toluene	110.6		79.5	99.95	0.06
Water	100	76.5	15.3	–	–
Acetonitrile	82		83.5	–	–
Benzene	81	64.5	74.1	86.0	4.8
Ethanol	78.5		18.5	12.7	52.1
Water	100		7.4	1.3	43.1
Toluene	110.6	74.4	51.0	81.3	24.5
ethanol	78.5		37.0	15.6	54.8
Water	100		12.0	3.1	20.7
Hexane	69.0	56.0	85.0	96.5	6.0
Ethanol	78.5		12.0	3.0	75.0
Water	100		3.0	0.5	19.0
Cyclohexane	80.7	62.5	75.5	–	–
Ethanol	78.5		29.7	–	–
Water	100		4.8	–	–

The fact that toluene (as well as acetonitrile, dichloromethane, *etc.*) on distillation will carry with it a certain percentage of water, it can be used to remove water (e.g. from a reaction mixture [azeotropic removal of water]).

Bibliography

[1] B. S. Furniss, A. J. Hannaford, P. W. G. Smith, and A. P. Tatchell, *Vogel's Texbook of Practical Organic Chemistry*, 5th edition, Addison Wesley Longman Ltd., 1989.

[2] *Technique of Organic Chemistry*, A. Weissberger ed., Vol. 4, 2nd edition, 1965.

[3] J. Benitez. *Principles and Modern Applications of Mass Transfer Operations*, John Wiley & Sons, Inc., New York, 2002.

[4] *CRC Handbook of Chemistry and Physics*, CRC Press, 86th editiom, 2005, Sect. D 1.

Chapter 5

Chromatographic Methods

5.1 BASIC PRINCIPLES

Chromatographic separation of mixtures of substances depends on the differences in the partition coefficients of the components between two immiscible phases [1]–[5]. One, the mobile phase, moves relative to the other, the stationary phase, and transports the substances being separated.

The partition coefficient K of a substance in such a two-phase system is given by

$$K = c_s/c_m,$$

where c_s = concentration of the substance in the stationary phase
c_m = concentration of the substance in the mobile phase

The higher the partition coefficient of a substance, the greater its concentration in the stationary phase and the slower is its movement along the chromatographic system.

According to the mechanism of retention one can further differentiate chromatography into the following types:

Adsorption (normal, reversed phase) chromatography
Liquid–liquid partition chromatography
Ion exchange chromatography
Gel permeation (exclusion) chromatography
Affinity chromatography

Practical Organic Synthesis: A Student's Guide   R. Keese, M.P. Brändle and T.P. Toube
© 2006 John Wiley & Sons, Ltd.

CLASSIFICATION

Mobile phase	Stationary phase	Chromatographic technique
vapour	solid	gas chromatography (gas–solid chromatography)
	liquid	gas chromatography (gas–liquid chromatography, GLC)
liquid	solid	adsorption chromatography (liquid–solid partition, liquid–solid chromatography, LSC)
	liquid	liquid–liquid partition (partition chromatography, liquid–liquid chromatography, LLC)

The following sections are devoted to partition between mobile liquid and stationary solid phases.

5.1.1 Adsorption Chromatography

Used for partition between mobile liquid and stationary solid phases.

The success of such separations depends mainly on the choice of the correct phases. The phases may be divided into:

(a) *Normal phases*
    stationary phase: polar (kieselgel, alumina, cellulose, *etc.*)
    mobile phase: nonpolar → polar (hexane → ether → methanol)
(b) *Reversed phases*
    stationary phase: nonpolar (modified kieselgels, nylon, polystyrene, *etc.*)
    mobile phase: polar (methanol/water/acetonitrile)

Stationary Phases

Kieselgel
Is by far the most common stationary phase used by preparative organic chemists to separate mixtures. Kieselgel is a dehydrated, highly porous silicic acid, ground to give a particle size of 0.04–0.2 mm, with pore diameters of 50–100 Å and a surface area of 200–400 m^2/g. The SiOH groups on the surface generally give kieselgel a weakly acid character, but different methods of preparation can be used to prepare acidic, neutral

or basic kieselgels. (They can be tested by suspending in water and measuring the pH; note that water is often slightly acidic because of dissolved CO_2.) Acidic kieselgel is not suitable for separating acid-labile substances.

Alumina
It is somewhat basic (pH 9.5). Neutral alumina is prepared by neutralisation to pH 7.0 followed by activation. Alumina can be deactivated by the addition of water.

Activity	I	II	III	IV	V
Water content (% by weight)	0	3	6	10	15

Preparation: Add the calculated amount of water. Shake until all lumps and damp patches disappear. Allow to stand *ca.* 24 hours in a tightly stoppered vessel.

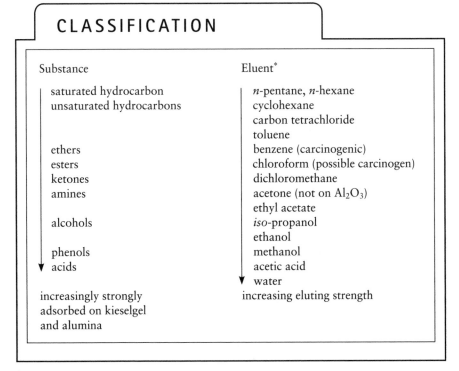

CLASSIFICATION

Substance	Eluent*
saturated hydrocarbon	*n*-pentane, *n*-hexane
unsaturated hydrocarbons	cyclohexane
	carbon tetrachloride
	toluene
ethers	benzene (carcinogenic)
esters	chloroform (possible carcinogen)
ketones	dichloromethane
amines	acetone (not on Al_2O_3)
	ethyl acetate
alcohols	*iso*-propanol
	ethanol
phenols	methanol
acids	acetic acid
	water
increasingly strongly adsorbed on kieselgel and alumina	increasing eluting strength

*Essential to evaluate any health hazards (see Chapter 14)

Most homogeneous mixtures of solvents can be used as eluents.

Only use solvents of known purity; they should normally be freshly distilled (ethers form peroxides, chloroform may contain alcohol or phosgene, solvents pick up moisture, *etc.*).

Temperature dependence

The lower the temperature, the more strongly substances are adsorbed on the stationary phase: carry out chromatography in an area which is draught-free and not too hot. In favourable cases, cooled columns may give improved separation, and may be advantageous for the separation of thermally labile compounds.

For inverse-phase systems, see the Bibliography.

5.2 ANALYTICAL AND PREPARATIVE THIN-LAYER CHROMATOGRAPHY [6]-[9]

5.2.1 Thin-layer Chromatography (tlc)

(a) *Uses*

Checking purity.
- Preliminary tests before separation.
- Qualitative comparison with known substances.
- A check on reactions.

(b) *Procedure*

(i) Cut commercial glass plates (20 cm × 20 cm) with a glass cutter (wheel, rather than diamond).

(ii) Choose stationary phase – alumina or kieselgel of *ca.* 0.2 mm thickness.

(iii) Apply the substance (or mixture) – about 2 μl of a dilute (*ca.* 1%) solution of the substance in the least polar suitable low-boiling solvent. Apply to the plate in the form of a spot and the solvent should be allowed to evaporate completely.

(iv) Develop the chromatogram. The plate is dipped into the eluent and allowed to develop. When the solvent front has advanced a suitable distance, the plate is removed (see Figure below), the solvent front marked, and the plate allowed to dry.

(v) Detection – coloured spots are, of course, immediately visible.

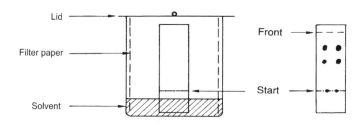

Colourless spots can be made visible by:

- UV: if the substance absorbs UV, one can use a stationary phase impregnated with a fluorescent indicator.
- standing the plate in iodine vapour, which detects many colourless compounds.
- spraying with conc. H_2SO_4–water 1:1 (in a specially protected compartment in a fume hood!!) and then heating strongly, *e.g.* with a hot-air blower, to carbonise the compounds.
- spraying with suitable colour reagents.[*]
  - 1% Phosphomolybdic acid in ethanol (limited lifetime): detects unsaturated and reducing compounds.
  - Alkaline $KMnO_4$ (0.5 g in 100 ml 1 M NaOH): detects unsaturated compounds and sugars.
  - Vanillin (1 g in 20 ml of EtOH with 80 ml of conc. H_2SO_4 added): detects higher alcohols, steroids, unsaturated hydrocarbons.
  - Bromocresol Green (0.5% in 80% MeOH, with eight drops of 30% NaOH added per 100 ml of solution): detects organic and inorganic acids.
  - Ninhydrin (0.3 g in 100 ml of butanol and 3 ml of glacial acetic acid): detects amino acids, primary amines and amino sugars).

(c) *Recording*
Trace the chromatogram using tracing paper. Mark in spots, starting position and solvent front and record type of plate, eluent and method of development.

R_f = (Distance of centre of spot from start)/(Distance of solvent front from start)

The R_f value depends on the conditions under which the chromatogram was run (type of plate, eluent, *etc.*, as well as condition of plate, temperature, saturation of vapour, *etc.*). Its reproducibility is about ±20%. It is best to run probable reference compounds on the same plate.

(d) *Multiple Runs*
It is sometimes useful to rerun a plate, after it has dried, until the solvent front reaches the position of the solvent front on the previous run. An *n*-fold rerunning is effectively the same as running a chromatogram *n* times the distance, and gives better resolution for very close spots of low R_f.

5.2.2 Preparative TLC (thick-layer chromatography)

(a) *Uses*
Preparative separation of mixtures (up to *ca.* 200 mg practicable).

(b) *Procedure*

---

[*] For a full list, see {"http://www.machery-nagel.de"} → chromatography/solid phase extraction → TLC → spray reagents; also [8] and *Dyeing Reagents for Thin Layer and Paper Chromatography*, E. Merck, Darmstadt, 1971.

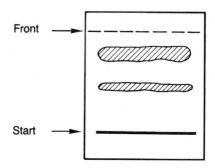

- By TLC optimise the system (adsorbent, eluent) for separation.
- Apply a concentrated solution (*ca.* 100 mg/ml) of the mixture to the plate as the narrowest possible strip, using a drawn-out pipette.
- Elute as for tlc
- Using a UV lamp if necessary, locate the individual bands and mark them with a pencil. (Alternative methods: cover the plate carefully with aluminium foil, leaving a 0.5 cm strip in the middle exposed, develop in iodine vapour and then assume the bands are straight! Especially if plastic or aluminium-backed plates are being used, a narrow marker strip can be cut from the centre and the bands located by one of the methods discussed in the section on analytical tlc).
- Scrape off the individual bands.
- Elute the substance with a solvent more polar than the eluent. *Warning:* Kieselgel is slightly soluble in ethanol or methanol. If one of these solvents has to be used, evaporate the alcohol solution after filtration, take up the residue in a less polar solvent and filter again.
- Evaporate the solvent. Crystallise or distil the product and characterise it.

Note: Many plates are slightly contaminated by plasticisers or binding agents. For small-scale work, plates should be cleaned by first running in a mixture of chloroform and methanol and then drying thoroughly before the mixture is applied. It is also essential to use pure, redistilled solvents for all processes.

5.3 COLUMN CHROMATOGRAPHY

(a) *Uses*
Separation of larger quantities of material (e.g. > 100 mg).

(b) *Procedure*
- By TLC optimise the system (adsorbent, eluent) for separation: The R_f value of the highest spot should not exceed 0.3.
- Quantity of adsorbent required: 30–100 times the weight of the mixture.

- Column dimensions: The adsorbent should pack into it with a 10:1 ratio of height:diameter. To estimate the quantity of adsorbent in relation to the column diameter, see pp. 46 and 47.
- Securely clamp column vertical.
- Fill the column with the eluent (or, if a mixed solvent is being used, with the less polar component). For choice of solvents, see Section 3(c).
- Press a plug of cotton wool firmly into the bottom with a glass rod. Cover it with a layer of clean sea sand (*ca.* 0.5–1 cm).
- Pour in the weighed absorbent:
  - Alumina – in a fine stream while tapping the column with a piece of rubber tubing.
  - Silica gel – as a slurry in the solvent.
- Allow the solvent to run out (tapping the column) until *ca.* 1 mm remains above the top of the adsorbent.
- Dissolve the substance in the minimum amount of solvent (not more polar than the eluent) and apply carefully and evenly to the top of the column (tap closed). Open the tap and run off solvent until a 1 mm layer again remains above the adsorbent. Repeat the process a few times using small amounts of eluent.

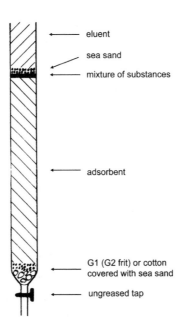

eluent

sea sand

mixture of substances

adsorbent

G1 (G2 frit) or cotton covered with sea sand

ungreased tap

- Fill the volume above the column of adsorbent carefully with eluent. To protect the surface of the adsorbent, a 1 cm layer of sea sand can be deposited on the top of the column. For choice of solvent, see Section 3(c).
- Open the tap and collect the eluent in fractions (ca. x ml fractions for x g of absorbent).
- Flow rate: poor separation results at high flow rates. Rule of thumb: 3–4 ml/min with a column 40 cm high.
- Evaporate each fraction and weigh the residue. Enter the information on the record (see below).
- Fractions containing the same components (check by tlc) can be combined, purified (crystallisation, distillation) and characterised.

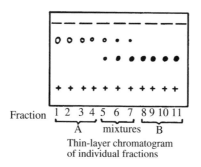

Thin-layer chromatogram
of individual fractions

(c) *Most Common Eluents (for Kieselgel and Alumina Columns)*
The choice of eluent depends on the polarity of the substances to be separated. One often needs to use mixtures of solvents. One often begins with petroleum. Dichloromethane and ethyl acetate are also good solvents for many substances.

COMMON ELUENTS

petroleum ether
toluene (rather than benzene)
diethyl ether
t-butyl ether

dichloromethane
ethyl acetate
methanol

The minimum quantity of a solvent mixture is used so that the solvent front reaches the end of the column before a mixture of different composition is applied.

USEFUL TIPS

- Do not let the column run dry.
- The temperature around the column should be kept approximately constant.
- A chromatographic run should be completed without interruption.
- When using alumina, do not have acetone as an eluent – you may find condensation products as the major constituents of the eluate.
- When low-boiling eluents are being used (ether, pentane, dichloromethane), especially when solvents are being changed, bubbles often appear in the column (channelling), and these usually prevent satisfactory separation being obtained. Use of cooled columns may prevent channelling.
- If it is anticipated that ether will be used as an eluent, it is advantageous to wash silica gel with a solvent mixture containing *ca.* 5% ether and then percolate a nonpolar solvent through the column before applying the substance to the column.
- For difficult separation problems the following procedure often works: dissolve the substance in the least polar solvent possible, add one to two times its weight of adsorbent, pump off the solvent and settle the dry powder onto the top of the column by pouring it through a small layer of eluent. Then elute as usual.
- For column chromatography it is often preferable to use an eluent mixture slightly less polar than the one shown to be the best for tlc.

(d) *Diagrams for Estimation of Column Dimensions and Quantities of Adsorbent for Column Chromatography*
The following diagrams enable one to estimate the amounts of silica gel (Figure 5.1) or alumina (Figure 5.2) required to fill a chromatographic column to a particular height.

EXAMPLE A

How much silica gel is required to fill about 4/5 of a column of diameter 18 mm and length 57 cm?

On Figure 5.1: for an 18 mm diameter, 26 g of silica gel (packed as a suspension) will give 18 cm diameter: height 10:1) of packing.
Therefore, for 45.6 cm (4/5 of 57) one requires 66 g ($45.6/18 \times 26$) of silica gel.

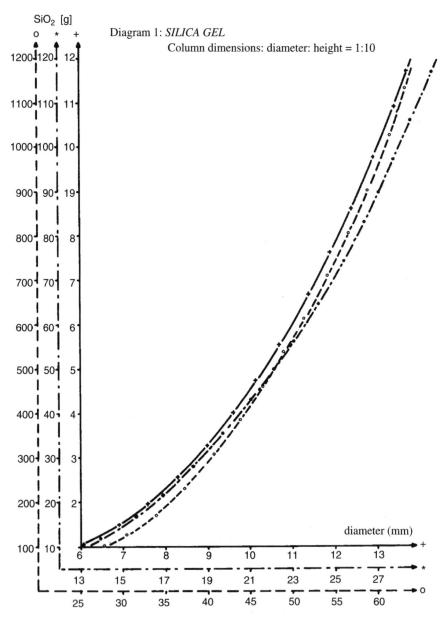

Figure 5.1: Diagram for the estimation of the quantity of silicagel required to give a column of adsorbent of diameter: height = 1:10

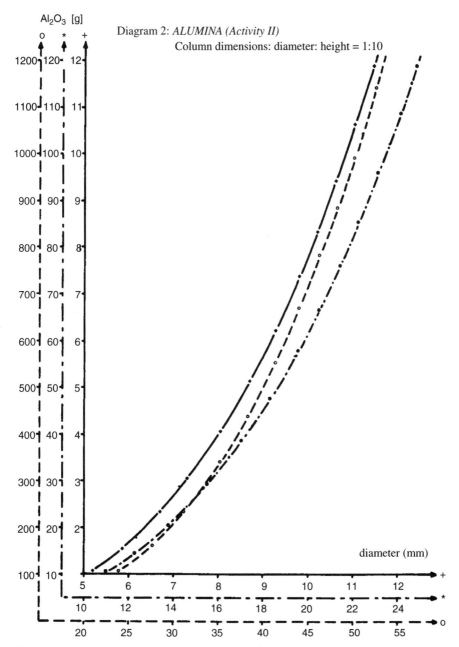

Figure 5.2: Diagram for the estimation of the quantity of alumina required to give a column of adsorbent of diameter: height = 1:10

CHROMATOGRAM

No. AM-7   Start: 09.30   Finish: 11.30
Substance: Reaction mixture   Weight: . . . g
Alumina (Woelm) a (n) b / Activity (III)⎱
                                            ⎰   10 g   55 fold amount
Kieselgel (. . .) a n / Mesh (   )
Adsorbed from: dichloromethane 1 cm^3
Column dimensions: diameter: 1.5 cm height: 17 cm

Fraction	Solvent	Ratio	Vol (cm^3)	Flask (g)	Wt (mg)	Total wt (mg)	Analytical data etc.
1	hexane/	1:1	10	66.345	0	0	
2	CH$_2$Cl$_2$	1:1	10	47.407	0	0	
3	"	1:1	10	52.578	0	0	
4	"	1:1	10	38.598	1.8	1.8	Combined:
5	"	1:1	10	36.189	35.1	36.9	tlc 17
6	"	1:1	10	39.857	17.5	54.4	IR 12
7	"	1:1	10	47.425	5.2	59.6	UV 5
8	"	1:1	10	44.577	1.2	60.8	
9	"	1:1	10	46.071	0	60.8	
10	"	1:5	20	35.346	0	60.8	
11	"	1:5	20	50.634	0	60.8	
12	"	1:5	30	45.514	0	60.8	
13	CH$_2$Cl$_2$		10	62.080	5.0	65.8	Combined:
14	"		10	52.578	11.2	77.0	Alc 18
15	"		10	59.167	31.3	108.1	IR 13
16	"		10	54.628	3.5	111.6	UV 6
17	"		10	42.407	2.1	113.7	
18	CH$_2$Cl$_2$/	1:1	20	53.212	3.2	116.9	
19	EtOAc	1:3	20	42.343	5.7	122.6	
20	"	1:3	50	36.189	4.0	126.6	
21	"	1:3	50	33.432	1.6	128.2	
22							

Table 5.1: Record of a chromatographic run

EXAMPLE B

A mixture requires 110 g of alumina for chromatographic separation. How large a column is required?

On Figure 5.2: 110 g alumina requires a column, diameter 24 mm and height > 24 cm. If one wishes to use a column whose diameter is 17 mm, calculate the height as in Example A. [110 g alumina will fill 49 cm].

5.4   RAPID COLUMN CHROMATOGRAPHY

5.4.1   'Flash Chromatography' [10]

(a) *Application*
   Separation of 100 mg – 10 g of mixtures of substances
   – components differ in R_f on tlc ≥ 0.15

(b) *Procedure*
   - By thin-layer chromatography (kieselgel) select an eluent which gives the R_f of the desired components in the range 0.2–0.3. If several components are to be isolated, the highest R_f should be 0.3.
   - Use a column with height:diameter ratio = 10:1, fitted with a pressure head and with a small dead-volume at the bottom.
   - Pack the column *dry* with kieselgel (40–63 μm mesh) 100–200 times the weight of the mixture to be separated. The column should be no more than 50% full.

- To protect the surface of the adsorbent and to improve separation between different solvents, add a 1-cm layer of fine sand.
- Fill the column with solvent and force it through the kieselgel using *ca.* 0.3 atm. pressure of an inert gas (N$_2$, Ar). Adjust the valve to give a flow rate of *ca.* 5 ml/min. *Care! Protective spectacles are essential!*
- Dissolve the mixture in a little solvent, apply to the column and allow it to percolate through the sand layer.
- Fill the column with eluent.
- Replace the head and apply the gas pressure.
- Collect fractions, 30 s – 1 min each.
- Evaporate fractions, weigh and test for homogeneity by tlc.
- Combine identical homogeneous fractions and characterise.

(c) *Practical Tips*

- For safety, cover the column with protective (polypropylene) mesh.

- Any column with B29 (or B34 or B45) joints and a small dead-space can be used for flash chromatography, if fitted with a commercial polypropylene stopper with two connecting nozzles as pressure head: The flow rate is regulated by means of a piece of tubing fitted with a clamp. It is useful to have a ball-and-socket ground glass joint (diam. 12.7 mm, bore 2 mm) at the foot of the column: the column is shut off by moving the socket to the side.
- If ether is used as eluent, the heat of interaction with the silica gel causes bubbles to appear in the adsorbent, which are difficult to eliminate. In this case it is better to pack the column with a slurry of silica gel.

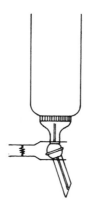

- Repeated use of the same column: after use, the column can be washed and then reused. First allow two column volumes of methanol to flow through, and then similar volumes of ethyl acetate and hexane (or the solvent mixture best suited to the next chromatogram). The retention time (or 'activity' of the column) is only slightly reduced by this procedure.
- Solvent can be recovered for re-use by distilling it, under reduced pressure if appropriate.

5.5 HPLC

The acronym stands for high-pressure liquid chromatography or high-performance liquid chromatography [11]–[14].

This efficient separatory technique has been very widely used since 1970. It can be used for analytical and preparative separations on a small scale (μg to *ca.* 200 mg). The high efficiency is achieved by the use of columns tightly packed with stationary phases of very fine, uniform mesh (typically 3–10 μm diameter) and with no dead volume. The columns are packed with a slurry of stationary phase at a pressure of 300–600 bar [10]. Compared to conventional column chromatography, HPLC requires more expensive apparatus and, as with gas chromatography, there is now a variety of HPLC apparatus available, essentially comprising pumps, columns, inlet systems, detectors and recorders. See Fig. 5.3.

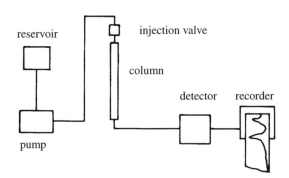

Figure 5.3: Schematic diagram of a HPLC apparatus

In addition, one can obtain a variety of stationary phases: various kieselgels (*e.g.* 5 ± 2 μm mesh), modified kieselgels, alumina and materials for reversed-phase HPLC. The most common detectors employ refractometric or spectrophotometric techniques.

The choice of particular separation systems depends on the nature of the substances to be separated. The following table may serve as a useful guide.

Class of substance	Separation method
Lipophiles, neutral organic compounds with MW < 2000	normal adsorption
hydrophiles, polar organic compounds, saturated hydrocarbons, MW < 5000	reversed-phase adsorption
moderately polar organic compounds	(liquid–liquid) partition
acids, bases, zwitterions, MW < 10 000	ion exchange
separation by molecular size of all sorts of compounds	gel permeation

In the following section, we assume that an efficient HPLC with a packed kieselgel column and fitted with a detector is available. A mixture of neutral organic substances is to be separated.

(a) *Procedure:*
- Determine the UV/visible spectrum of the mixture.
- Consider only those solvents transparent in the wavelength regions where the compounds absorb. The most useful solvents[**] for HPLC are hexane (200), di-ethyl ether (220), *t*-butyl methyl ether (220), dichloromethane (230), ethyl acetate (253), chloroform (242), acetonitrile (200), methanol (210) and water.
- Use TLC to choose the best solvent (or mixture of solvents). The R_f values for the substances to be separated should be ≤ 0.3. If the R_f reaches 0.9–1 even in hexane, kieselgel is not a suitable stationary phase; reversed-phase HPLC may be considered.
- To obtain consistently high separation efficiency the chosen solvent should be prepared as follows:
  - (i) Use solvents only of known purity, HPLC quality if necessary.
  - (ii) To obtain a defined water content in the mobile phase, mix solvent saturated with water and absolute solvent in a known proportion (*e.g.* 1:1).
  - (iii) Remove any dust particles from the solvent by filtration through a sinter (≤1 μm pore size).
  - (iv) De-gas the solvent: apply moderately reduced pressure or use ultrasonics.
- Condition the column with the chosen solvent (or mixture) until the detector and recorder show a constant baseline. Large, sudden changes in solvent polarity should be avoided.
- Apply a small quantity (*ca.* 0.1 mg) of the mixture, dissolved in the eluting solvent, to obtain a test chromatogram. Adjust the polarity of the mobile phase to obtain a reasonable retention (k' *ca.* 3). If no separation results, alter the polarity to increase k' (to *ca.* 10). If there is still no separation, change to a mobile phase composed of other solvents.

[**] The figure is the wavelength in nanometres at which the transmittance of a 1-cm layer of the solvent is < 10%.

- For reliable separations, one must ensure that the capacity factor k' and the resolution R_s (see p. 54) remain constant each time the substance is applied.
- Collect fractions giving the same signal.
- After concentration of each component, test for purity by applying once more.

For separation on reversed-phase columns, water–methanol, water–acetonitrile or water–tetrahydrofuran mixtures are used. If the apparatus has no facilities for gradient elution, test first using a 1:1 solvent mixture. If k' is too low, increase the proportion of water; if the substance is not eluted at all (*i.e.* k' is too large), increase the proportion of the organic solvent.

(b) *Some additional concepts, plus the calculation of the number of theoretical plates of a column* (including connections and detector cell):

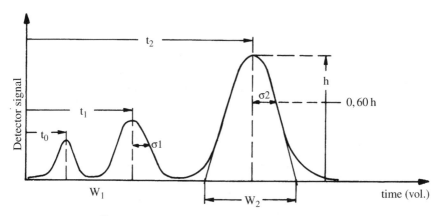

Figure 5.4: Sketch of a HPLC chromatogram

L = column length
t_0 = time between injection and the arrival at the detector of a non-adsorbed substance ('solvent front') [*e.g.* CCl_4 would be carried at the solvent front using hexane as eluent]
t_1, t_2 = time taken before the first and second substances produce a signal at the detector
h = signal height
σ = half-width of the signal at 0.60 of its height
w = signal width, as given by its side tangents
N = no. of theoretical plates (based on the 1st signal)
$$N = (t_1/\sigma_1)^2 = 16(t_1/w_1)^2$$
H = height (equivalent) per theoretical plate
$$H = L/N$$
k' = capacity factor (which is a measure of the retention of a substance, *cf.*, the R_f value in tlc)

$$k'_1 = (t_1 - t_0)/t_0$$

(k' has values ≥ 1, except in exclusion chromatography)

R_s = resolution of two signals

$$R_s = (t_2 - t_1)/[^1\!/_2(w_1 + w_2)]$$

References

[1] J. M. Miller. *Chromatography: Concepts and Contrasts*, 2nd edition, Wiley-InterScience, 2004.

[2] L. R. Snyder, J. J. Kirkland. *Introduction to Modern Liquid Chromatography*, 2nd. edition, Wiley-InterScience, 1979.

[3] C. F. Poole. *The Essence of Chromatography*, Elsevier Science, 2002.

[4] P. Millner. *High Resolution Chromatography: A Practical Approach*, Oxford University Press, Oxford, 1999.

[5] K. Hostettmann, A. Marston, and M. Hostettmann. *Preparative Chromatography Techniques, Applications in Natural Product Isolation*, 2nd edition, Springer Verlag, 1998.

[6] P. E. Wall. *Thin Layer Chromatography. A Modern Practical Approach*, Royal Society of Chemistry, 2005.

[7] J. Sherma and B. Fried. *Handbook of Thin-Layer Chromatography*, 2nd edition, Marcel Dekker, New York, 2003.

[8] E. Hahn-Deinstrop. *Applied Thin Layer Chromatography: Best Practice and Avoidance of Mistakes*, Wiley, 2000.

[9] H. Jork. *Thin-Layer Chromatography: Reagents and Detection Methods*, Wiley, 2006.

[10] W. C. Still., M. Kahn, and A. Mitra. *J. Org. Chem.*, 1978, 43, 2923.

[11] V. R. Meyer. *Practical High-Performance Liquid Chromatography*, 4th edition, Wiley, 2004.

[12] L. R. Snyder, J. J. Kirkland, and J. L. Glajch. *Practical HPLC Development*, 2nd edition, Wiley-Interscience, New York, 1997.

[13] E. Pritchard. *High Performance Liquid Chromatography (Practical Skills Training Guide)*, Royal Society of Chemistry, London, 2003.

[14] P. C. Sadek. *The HPLC Solvent Guide*, 2nd edition, John Wiley & Sons, 2002.

Chapter 6

Extraction and Isolation

Mixtures, especially those from reactions, can often be separated at little expense by extraction [1].

Extraction can be defined as the transfer of a substance X from one liquid phase A into an immiscible liquid phase B. The partition of X between the immiscible liquids

Practical Organic Synthesis: A Student's Guide   R. Keese, M.P. Brändle and T.P. Toube
© 2006 John Wiley & Sons, Ltd.

A and B is given by the Nernst partition equation.

$$C_B(X)/C_A(X) = K_T$$

where $C_A(X)$ and $C_B(X)$ are the concentrations of X in A and B, respectively.
K_T = the partition coefficient at temperature T.

If X is much more soluble in B than in A ($K \geq 100$), two to three extractions will suffice to 'shake out' X. Conversion of a substance into its salt alters its partition coefficient drastically. By this means, organic molecules may be rendered water soluble and, as we see below, can then be separated easily from other (neutral) substances by shaking out.

6.1 SEPARATION OF MIXTURES ACCORDING TO ACIDITY

This variant of extraction is one of the commonest processes in preparative organic chemistry. The following procedure is, therefore, particularly suited to preparative experiments.

6.1.1 Principle

Carboxylic acids and phenols are deprotonated by bases (sodium bicarbonate, and sodium carbonate or hydroxide, respectively) and the anions are then water soluble. Amines are rendered water soluble by protonation in acidic media.

6.1.2 Advantages

Rapid separation and consequential safer handling of the relevant substances, especially as one often rinses them out using cold aqueous solutions (addition of ice).

6.1.3 Disadvantages

The solvents must be removed subsequently. Also, the method is not suitable for compounds that react with water.

6.1.4 Procedure

(a) *Preliminaries*
- Prepare sufficient freshly distilled ether.
- Prepare requisite aqueous solutions:
  - saturated $NaHCO_3$ (95.7 g/l, *ca.* 1 N).
  - saturated Na_2CO_3 (211.97 g/l, *ca.* 1 N).
  - 2 N NaOH (80 g/l).
  - 2 N HCl (200 ml/l conc. HCl).
  - saturated NaCl.

- Have H_3PO_4 ready for acidification of basic extracts (2 M H_2SO_4 as second choice).
- Moisten separatory funnel taps with water (or, if necessary, grease lightly); make sure they do not leak!
- Have a supply of ice available.
- Make sure there are no naked flames in the vicinity.

(b) *Separation*

- Dissolve mixture in ether. Filter if necessary (and in that case triturate the residue three times with ether, filter off the solvent and combine the ethereal extracts). The ether-insoluble material should be tested to see whether it dissolves in water.
- Put ice into three separatory funnels.
- Put some ether into the second and third funnels.
- Put the ethereal solution into the first funnel.
- During the following operations the ether layers remain in their respective funnels. Funnels should be no more than two-thirds full at any time.
- Hold the stoppers and taps in place while shaking. Shake carefully at first, release internal pressure via the tap (invert the funnel!) repeatedly until no excess pressure remains and then shake fairly vigorously about 20 times.
- The funnels should be at most two-thirds full, with a ratio of organic:aqueous phases of about 2:1.
- Add a portion of saturated $NaHCO_3$ solution to the first funnel; shake. Transfer aqueous layer to second funnel, shake; transfer to third funnel; shake. Run off the aqueous layer.
- Repeat the process with two more portions of saturated $NaHCO_3$. Combine the $NaHCO_3$ extracts.
- Now (as in Figure 6.1a) continue the extraction procedure in the same way, successively using the other aqueous solvents (*viz.* saturated Na_2CO_3, 2 N NaOH, 2 N HCl).
- In the final stage, extract the ether layers with saturated NaCl until the aqueous layer is neutral (it may be necessary to add a trace of $NaHCO_3$.)

(c) *Extraction of Organic Substances from the Aqueous Layers* (Figure 6.1b)

- Work up the basic layer, containing phenolic salts, first (they are the most likely to decompose).
- Each basic aqueous layer is mixed with ice and then carefully acidified to pH 1–2 in a large beaker using H_3PO_4 (*Careful*: CO_2 is evolved).
- Ice is added to the HCl extract which is then basified to pH $\approx$ 10 with 2 N NaOH.
- For the extraction of each of the four aqueous solutions, place them in equal portions in three separatory funnels, and shake with ice and ether.
- The three ethereal extracts from each solution are combined, dried (over anhydrous $MgSO_4$ except that the bases should be dried over anhydrous K_2CO_3 and acid-sensitive substances over Na_2SO_4), filtered, the filtrates evaporated and the residues dried to constant weight.
- The aqueous layers should be saved until one has accounted for all the material.

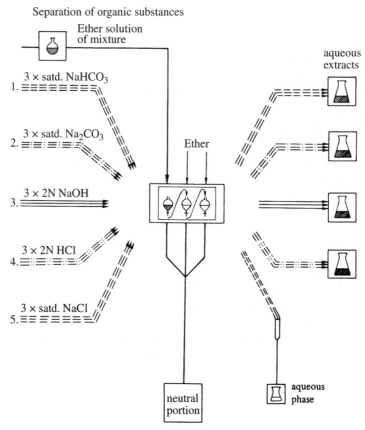

Separation of organic substances

○ Figure 6.1a: Scheme for the separation of organic substances according to acidity

(d) *Some Hints*
- Work quickly (separatory funnels are not for storing solutions!).
- Repeated extractions using small portions are more efficient than one extraction by a large portion.
- Label your solutions immediately!
- Emulsions are sometimes formed. They can often be broken by adding a few drops of methanol, amyl alcohol, *etc.* Otherwise try filtration or rotating the flask. Sometimes all you can do is wait!

6.1.5 Special Apparatus for Extractions

For polar, nonionic compounds, the difference in the partition coefficients between aqueous and organic phases may be so small that extracting three times with ether,

Extraction of organic substances from aqueous phase

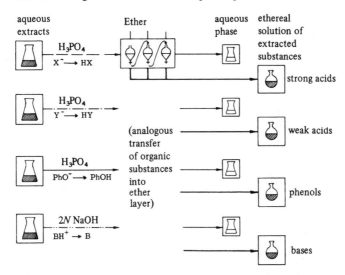

Figure 6.1b: Scheme for extraction of organic substances from the aqueous phase

dichloromethane or ethyl acetate may not be sufficient. In these cases one needs to use special apparatus for continuous solvent extraction. Apparatus A (below) is used for extraction with a solvent which is denser than the aqueous solution; apparatus B is used for a solvent less dense than the aqueous phase. A Soxhlet (apparatus C) can be used for continuous extraction of a solid.

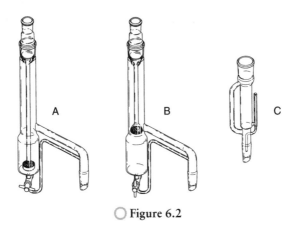

Figure 6.2

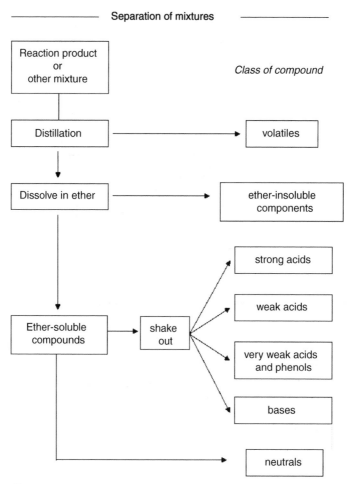

Figure 6.3a: Survey of methods for the separation of mixtures

——————————————— Purification and Identification ———————————————

Purification and further workup	Identification
Distillation	*Spectroscopic*
Sublimation	Ir, uv, ms, nmr (^{1}H, ^{13}C, etc.)
Crystallization	$[\alpha]_D$, ORD, CD
	Physical:
Preparative $\left.\begin{array}{l}\text{t.l.c.}\\\text{Column}\\\text{Chromatography}\end{array}\right\}$ + Distillation / Crystallization	Comparison with authentic material by:
	Mixed m.p.
	Mixed g.l.c.
Gas chromatography	Mixed t.l.c.
HPLC	HPLC mixed k'-value
Electrophoresis	Refractive index
Zone melting	pK
Dialysis	Density
Purification via a derivative	Heat of combustion
	Molecular weigtht
Criteria for purity	*Chemical:*
Constant m.p.	Via derivatives
One t.l.c. spot in several systems	Elemental analysis
Single g.l.c. peak on a variety of columns	

○ Figure 6.3b: Survey of methods for the purification and identification of organic substances

Bibliography

[1] J. Rydberg, M. Cox, C. Musikas and G. R. Choppin. *Solvent Extraction Principles and Practice*, 2nd edition., Marcel Dekker, 2004.

[2] *Techniques of Organic Chemistry*, A. Weissberger, Ed., Vol. 3, p. 171, Inter-Science, 1950; *Techniques in Chemistry*, 3rd edition., Vol. 12, John Wiley & Sons, 1978.

Chapter 7

Structure Determination Using Spectroscopic Methods

Instrumental methods are routinely used these days to assign structural features and functional groups to organic molecules from the positions, intensities and patterns of spectroscopic signals. These data may be used to determine the structures of unknown compounds, to confirm the structures of synthetic products or to provide information concerning the progress of a reaction. In this section we illustrate, by means of a simple example, how a combination of common spectroscopic techniques (IR, MS, ^{1}H and ^{13}C NMR and UV) can be used to determine the structure of a compound.

All recent textbooks include chapters on spectroscopic methods in which the relationships between the types and positions of signals and the corresponding functional groups and structural features are explained, so that we provide here only a summary of the structural elements which may be identified spectroscopically. For a detailed discussion of the basis and scope of each technique, consult the specialist texts listed in the Bibliography at the end of this section.

7.1 INFRARED SPECTROSCOPY (IR)

Vibrations and deformations in the range 4000–200 cm^{-1} are characteristic of many functional groups.

Practical Organic Synthesis: A Student's Guide   R. Keese, M.P. Brändle and T.P. Toube
© 2006 John Wiley & Sons, Ltd.

7.2 MASS SPECTROMETRY (MS)

The radical cation M^+ derived from a molecule M in the vapour phase gives information about the relative molecular mass and – at high resolution – molecular formula. The molecular ion (M^+) is defined as the ion whose mass corresponds to the molecular composition of the compound, taking the most abundant isotope in each case.

In determining elemental composition it is important to remember that M^+ is always even provided the molecule contains only atoms whose valencies and atomic weights are either both even or both odd (H, C, O, S, Si, Cl). For other elements or isotopes, such as ^{14}N, ^{13}C and ^2H, the value of M^+ is odd unless the number of such atoms is even. In addition, fragmentation of M^+ must involve the loss of only chemically reasonable fragments. Accordingly, mass losses of 3–13, 21–24, 37 and 38 are not reasonable.

Besides the molecular ion, peaks are observed that derive from natural isotopic abundances. From the (M + 1) peak one may estimate the maximum number of C atoms; (M + 2) is diagnostic of *e.g.* Cl, Br, Si and S.

7.3 NUCLEAR MAGNETIC RESONANCE SPECTROSCOPY (NMR)

Certain nuclei, such as ^1H, ^{13}C, ^{19}F and ^{31}P, possess a magnetic moment which can adopt parallel or antiparallel orientations in a magnetic field. Transitions between the energy levels associated with these two orientations are induced by radio frequency electromagnetic radiation. The measured energy difference depends, among other factors, on the chemical environment of the particular nucleus, and from this one can derive a 'chemical shift' which is characteristic of the environment of the nucleus. For ^1H and ^{13}C, this chemical shift δ [ppm] is measured relative to tetramethylsilane (TMS). In ^1H NMR the area under a peak is proportional to the number of protons having the same chemical shift.

Nuclei possessing a magnetic moment, which are close to one another in a molecule, interact to produce a mutual coupling that may lead to the spin–spin splitting of the signals at the relevant chemical shifts. The pattern of coupling and the size of the coupling constant, J (Hz), contain substantial structural information. For coupling between two nuclei, A and B, $J_{AB} = J_{BA}$. The structural information from ^1H NMR spectra is considerably amplified and extended by ^{13}C NMR spectroscopy. One distinguishes between 'broad-band decoupled' ^{13}C spectra, which give only the chemical shifts of the ^{13}C atoms, and 'off-resonance' spectra, which retain residual couplings from ^1H nuclei bonded directly to the ^{13}C atoms.

7.4 ULTRAVIOLET AND VISIBLE SPECTROSCOPY (UV/VIS)

Band positions and intensities for electronic transitions supply information concerning the type and extent of conjugated IR systems.

The determination of the structure of an unknown compound rests on the amalgamation of structural elements derived from spectroscopic data into the most probable combination.

The flow diagram below summarises the procedure.

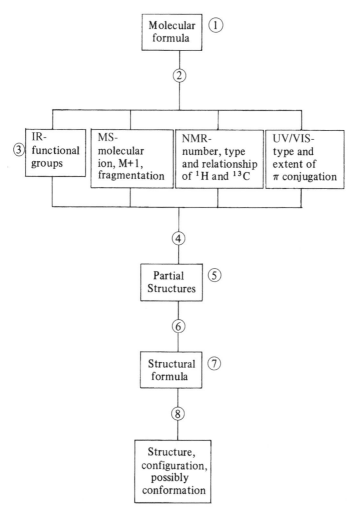

◯ Figure 7.1: Scheme for determining a structure from spectroscopic data

(1) Elucidation of a structure generally starts with the molecular formula. This may be determined from the empirical formula from combustion analysis plus the molecular weight. The latter may be obtained either from the molecular ion (M^+) in the mass spectrum or by other techniques.

Once the molecular formula is known, calculate F, the number of double-bond equivalents (double bonds + rings):

$$F = 1 + \frac{\Sigma(\text{no. of atoms of a particular element}) \cdot (\text{formal valency} - 2)}{2}$$

In the simpler case of compounds $C_xH_yO_z$, this formula reduces to:

$$F = 1 + (x - y/2)$$

In conjunction with the UV/VIS spectrum, F provides information on the number of double bonds and rings.

(2) From the spectra deduce probable structural elements; check whether they are also detectable in the other spectra. In addition, note which functional groups are absent, especially from the IR.

(3) IR: Analyse particularly the regions 4000–3100, 3100–2700 and 1800–1650 cm^{-1}.

MS: The peak of highest m/z is often, but not always, M$^+$. Typical fragmentation patterns are known for many classes of compounds.

NMR: Tables of chemical shifts and coupling patterns can often lead to unambiguous partial structures.

UV/VIS: Tables can lead to the identification of many chromophore structures.

(4) Keep a running check between the structural elements determined according to (3) and the molecular formula in order to decide how many atoms and groups are still missing or whether the sum of the partial structures exceeds the molecular formula.

(5) If a number of partial structures or structural elements are compatible with the molecular formula, consult additional analytical data, *e.g.* fingerprint region in IR, double resonance in NMR.

(6) Combine the partial structures, taking into account factors such as symmetry in the NMR spectrum, probable fragmentation in MS, *etc.* If it seems likely that the substance is known, consult tables of melting points, refractive indices, *etc.*, which may lead to a rapid, unambiguous confirmation of the proposed structure.

(7) The final structural proposal must be evaluated in terms of the expected spectral data (tables, additivity rules, spectra of model compounds). All the observed spectroscopic characteristics must be plausibly explicable. If the results do not lead to an unambiguous structure, one or more chemical reactions may lead to a compound whose structure may be determined uniquely by spectroscopic means.

(8) Configurations and, where appropriate, conformations are best deduced separately from the analytical data.

Worked example

Deduce the structure of the compound which gives the following data.

Elementary analysis

C 73.17% H 7.32%

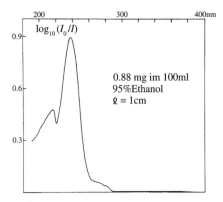

Figure 7.2a: Ultraviolet spectrum

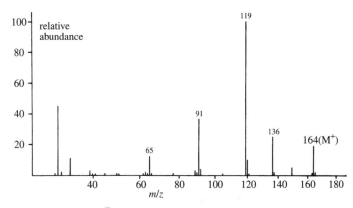

Figure 7.2b: Mass spectrum

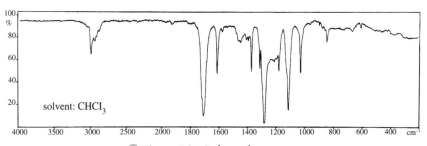

Figure 7.2c: Infra-red spectrum

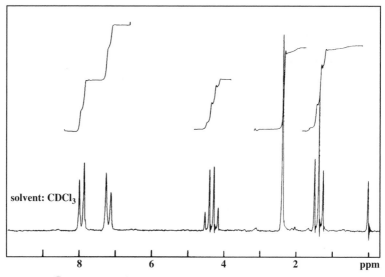

○ Figure 7.3: ¹H nuclear magnetic resonance spectrum

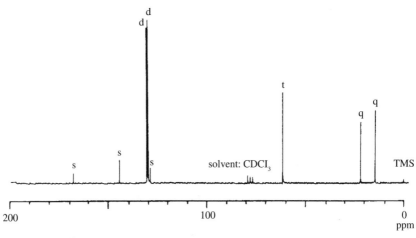

○ Figure 7.4: ¹³C nuclear magnetic resonance spectrum

Solution:

molecular formula:

C 73.17% H 7.32%; MS: $M^+ = 164$

$$C = \frac{0.7317 \times 164}{12} = 10.0 \quad H = \frac{0.0732 \times 164}{1} = 12.0$$

$C_{10}H_{12}\ldots = 132 + x = 164{:}x = 32$, *i.e.*, O_2 or S

MS: S would give M+2 *ca.* 4% (^{32}S:^{34}S = 95:4.2), therefore O_2, not S.

molecular formula $C_{10}H_{12}O_2$

$$F = 1 + \frac{10(4-2) + 12(1-2) + 2(2-2)}{2} = 5$$

UV:

$\nu_{max} = 237\,nm$

Calculation of c at 237 nm
$\varepsilon \times c \times l = \log I_0/I$

   ε = molar extinction coefficient
   c = concentration (mol l^{-1})
   l = path length (cm)

therefore $\varepsilon = \dfrac{\log I_0 I}{c \times l} = \dfrac{0.89}{1 \times 5.37 \times 10^{-5}} = 1.6586 \times 10^4$

ε suggests a conjugated system.

IR:

1705 cm^{-1}	C = O stretch (aryl or α,β-unsaturated ester, aryl aldehyde, saturated ketone)
1610 cm^{-1}	possible aromatic ring
1280 cm^{-1}	$-$C$-$O$-$stretch
~3000 cm^{-1}	CH stretch

No bands for $-$OH, $-CO_2H$ ($-$CN, $-CONH_2$).

MS:

(M + 1) *ca.* 10% of M$^+$, so 9–10 C atoms
Fragments

   164–149 (CH$_3$, 15)
   164-+136 (C$_2$H$_4$ or CO, 28)
   164–119 (OC$_2$H$_5$, 45)
   119–91 (CO, 28)

Weaker fragments:

m/z 39, 51–53, 63–65, 77–8: aromatic ring fragments.

NMR:

^{1}H: Intensities: 40:40:40:60:60→2:2:2:3:3
        Σ = 12, consistent with molecular formula

Chemical shifts δ [ppm]	Assignment
1.35 (t, J 7Hz, 3H)	$-CH_2-CH_3$
2.40 (s, 3H)	$-C-CH_3$
4.32 (q, J 7Hz, 2H)	$-O-CH_2-CH_3$
7.18 (2H)⎤	AA' BB' system
7.90 (2H)⎦	

- The signals at δ = 1.35 and 4.32 are coupled together, with pattern for $-CH_2CH_3$ and chemical shift of CH_2 suggesting $-OCH_2CH_3$.
- The signals at δ = 7.18 and 7.9 are in the aromatic region and the coupling pattern is a typical AA' BB' system for a 1,4-disubstituted benzene ring.

^{13}C (broad band decoupled)

Chemical shifts δ [ppm]	Multiplicity 'off-resonance' spectrum	Assignment
14.34	q (CH_3)	⎤ $^{13}CH_3$ not attached
21.55	q (CH_3)	⎦ to heteroatom
60.61	t (CH_2)	$^{13}CH_2-O-$
127.73	s ($-C-$)	⎤
128.84	d (CH)	aromatic
129.40	d (CH)	⎦ ^{13}C
143.15	s ($-C-$)	
166.40	s ($-C-$)	

At least eight C-atoms. From the molecular formula, two C-atoms must be duplicated (symmetry)→possibly a 1,4-disubstituted benzene

Partial structures: $-CH_3$, $-CO-O-$, $-OCH_2CH_3$, 1,4-disubstituted benzene

Proposed structure

This structure is compatible with

$F = 5$ (4 double bonds + 1 ring)
^{1}H coupling: $-CH_2-CH_3$; $^3J_{H-H}$ = 7Hz (cf. ethyl acetate)

^{13}C chemical shifts: The estimated chemical shifts (sequence according to numbering) are δ = 14.0, 21.5, 60.6, 125.6, 128.7, 129.9, 144, 165.7 [32] The reported chemical shifts (measured in $CDCl_3$) are δ = 19.93, 21.09, 60.24, 127.54, 128.70, 129.53, 142.91, 166.05 [33] methyl p-methylbenzoate has the following signals (sequence given according to the numbering):

21.4 (Ar–CH_3); 51.6 (OCH_3); 127.9, 129.2, 129.7, 143.4 (Ar); 166.8 (C = O).

MS: M^+ contains all the elements found in the fragments and by other spectroscopic methods. The fragments m/z 136 arises from the McLafferty rearrangement reaction.

Bibliography

[1] R. M. Silverstein, F. X. Webster and D. Kiemle. *Spectrometric Identification of Organic Compounds*, 7th edition, Wiley, 2004.

[2] E. Pretsch, P. Bühlmann and C. Affolter. *Structure Determination of Organic Compound: Tables of Spectral Data*, 3rd edition, Springer-Verlag, 2004.

[3] R. J. Anderson. *Organic Spectroscopic Analysis*, Royal Society of Chemistry, 2004.

[4] D. L. Pavia, G. M. Chapman and G. S. Kriz. *Introduction to Spectroscopy*, 3rd edition, Brooks Cole, 2000.

[5] E. Pretsch, G. Tóth, M. E. Munk and M. Badertscher. *Computer-Aided Structure Elucidation: Spectra Interpretation and Structure Generation*, Wiley-VCH, 2003.

[6] (a) R. L. Shriner, T. C. Morill, C. K. Hermann, D. Y. Curtin and R. C. Fuson. *Systematic Identification of Organic Compounds, Student Solutions Manual*, 8th edition, John Wiley & Sons, 2003. (b) C. K. F. Hermann and R. L. Shriner. *The Systematic Identification of Organic Compounds*, Int'l ed., 8th edition, John Wiley & Sons, 2003.

[7] Y.-C. Ning. *Structural Identification of Organic Compounds with Spectroscopic Techniques*, John Wiley & Sons, 2005.

[8] M. Hesse, H. Meier, B. Zeeh, A. Linden and M. Murray. *Spectroscopic Methods in Organic Chemistry*, Thieme Medical Publishers, 1997.

[9] J. B. Lambert, H. F. Shurvell, D. A. Lightner and R. G. Cooks. *Organic Structural Spectroscopy*, Prentice Hall, 1997.

[10] D. H. Williams and 1. Fleming. *Spectroscopic Methods in Organic Chemistry*, 5th edition, McGraw Hill, 1995.

[11] H. Günzler and H.-U. Gremlich. *IR Spectroscopy: An Introduction*, Wiley-VCH, 2002.

[12] B. H. Stuart. *Infrared Spectroscopy: Fundamentals and Applications*, John Wiley & Sons, Ltd., 2004.

[13] E. Breitmaier. *Structure Elucidation by NMR in Organic Chemistry: A Practical Guide*, 3rd edition, John Wiley & Sons, Ltd., 2002.

[14] S. Berger, S. Braun. *200 and More Basic NMR Experiments: A Practical Course*, John Wiley & Sons, 2004.

[15] R. S. Macomber. *A Complete Introduction to Modern NMR Spectroscopy*, John Wiley & Sons, 1997.

[16] H. Friebolin. *Basic One- and Two-Dimensional NMR Spectroscopy*, 4th edition, John Wiley & Sons, 2004.

[17] H.-O. Kalinowski, S. Berger and S. Braun. *Carbon 13 NMR Spectroscopy*, John Wiley & Sons, 1989.

[18] K. Pihlaja and E. Kleinpeter. *Carbon-13 NMR Chemical Shifts in Structure and Stereochemical Analysis*, Wiley-VCH, 1994.

[19] M. Balic. *Basic ^1H- and ^{13}C-NMR Spectroscopy*, Elsevier Science, 2005.

[20] P. J. Hore, J. A. Jones, S. Wimperis and S. Wimperis. *NMR: The Toolkit*, Oxford University Press, 2000.

[21] F. W. McLafferty and F. Tureček. *Interpretation of Mass Spectra*, 4th edition, University Science Books, 1993.

[22] R. M. Smith. *Understanding Mass Spectra: A Basic Approach*, 2nd edition, John Wiley & Sons, Ltd., 2004.

[23] E. De Hoffmann and V. Stroobant. *Mass Spectrometry: Principles and Applications*, 2nd edition, John Wiley & Sons, 2001.

[24] T. A. Lee, *A Beginner's Guide to Mass Spectral Interpretation*, Wiley, 1998.

[25] M. J. K. Thomas and D. J. Ando. *Ultraviolet and Visible Spectroscopy: Analytical Chemistry by Open Learning*, 2nd edition, John Wiley & Sons, Ltd., 1996.

[26] H.-H. Perkampus, H. C. Grinter and T. L. Threlfall. *Uv-Vis Spectroscopy and its Applications*, Springer, 1992.

[27] J. M. Hollas. *Modern Spectroscopy*, 4th edition, John Wiley & Sons, 2004.

[28] J. Workman Jr. and A. W. Springsteen. *Applied Spectroscopy*, Academic Press, 1998.

[29] **SpecTool**: Software for interpretation of NMR, MS and UV/VIS spectra, estimation of ^1H and ^{13}C NMR shifts and calculation of molecular formulae from spectra: www.upstream.ch.

[30] **ACD/CNMR** and **ACD/HNMR**: Integrated software package from Advanced Chemistry Development Inc. that calculates ^1H, ^{13}C spectra for many organic structures: www.acdlab.com.

[31] WebSpectra, Problems in NMR and DR Spectroscopy: http://www.chem.ucla.edu/~webspectra/.

[32] Program $^1H-$, $^{13}C-NMR$ predictor, available from ACD (Advanced Chemistry Development, Inc., Toronto, ADC version 9.08.

[33] B. A. Kellog, J. E. Tse, R. S. Brown, J. Am. Chem.. Soc., 1995, 117, 1732.

For information on chemical spectra and spectral data see the compilation given in Chapter 8.

Chapter 8

Searching the Chemical Literature

8.1 THE INFORMATION MOUNTAIN

The steadily increasing range of the chemical literature makes it increasingly difficult to find the information one requires. The number of papers published currently doubles roughly every 15 years. By March 2007, more than 89 million compounds had been described, of which about 58 million were nucleotide or peptide sequences.

Practical Organic Synthesis: A Student's Guide   R. Keese, M.P. Brändle and T.P. Toube
© 2006 John Wiley & Sons, Ltd.

Fortunately, there are now a number of effective computerised methods of searching data banks reliably.

The chemical literature can be divided into two main categories. The primary literature includes journal papers (there are currently some 9000 relevant journals), the patent literature, theses and dissertations, and conference and research reports. The secondary literature can be roughly subdivided into databases, like Beilstein, Gmelin, Chemical Abstracts and Web of Science, and review publications such as 'Science of Synthesis', 'Techniques of Chemistry' and 'Synthetic Methods of Organic Chemistry', as well as monographs and reports of 'Advances'. The secondary literature also includes publications such as information from chemical suppliers, as well as compilations of data on poisons and hazards, and safety regulations. One can also distinguish between databases that contain information about compounds and their properties and reactions, with appropriate literature references, and those which consist only of references from the primary literature.

The present Chapter deals with those databases which are most important for chemistry, illustrated by typical searches for information. There is also a bibliography of some essential printed reviews and collections of data.

8.2 STRATEGY

For an efficient search of the literature it is essential that one thinks carefully about defining exactly the purpose of the search. Only when the required data are described unambiguously will the process lead rapidly to the desired information.

To formulate a strategy it is useful to ask the following questions:

- What information is required? Is it clearly defined?
- How rapidly is it needed?
- Is a full search of the databases essential or will a preliminary survey suffice? Are only particular aspects of the compound or area of interest?
- Which databases and reviews are readily available nearby?
- At which point should one move from the databases to consulting the original literature?
- What role does serendipity play in finding information that could be of relevance in a different field?

Proceed from the simple to the complex: many of the works in the Bibliography, such as the handbooks, contain useful hints on searching databases.

8.3 DATABASES

The databases described in this section, CrossFire Beilstein [Elsevier Molecular Design Limited; www.mdl.com], SciFinder Scholar [www.cas.org/SCIFINDER/SCHOLAR/

index.html] of Chemical Abstracts [www.cas.org], and Web of Science [Thomson Scientific; www.isinet.com], are frequently available to institutions of higher education under licensing agreements that do not charge on the basis of time. This means that users have the opportunity to experiment with search strategies and thus really get to know their way around the databases. Nevertheless, it is essential to set out using a strategy with as much flexibility as possible in defining its scope, as experience shows that one often gets large numbers of hits at first. Modern software is so user-friendly that the underlying complex structure of the databases is completely hidden.

It is essential that one understands the structure of each database and its functions if one is to be able to conduct searches for scientific data reliably and efficiently. The following procedures are advisable:

- Begin searches for organic or metallo-organic compounds and their reactions with a structure. This is the only unambiguous way of defining a compound; if necessary, add the configuration as a refinement. Searching by name can lead to errors, as nomenclature is not sufficiently unambiguous and trivial names are often omitted in the database. For salts and solid-state compounds the situation is more complex: Consult the search information provided by the database.
- Check the results for relevance: Do they answer the question posed? Spot checks on the hits may provide useful hints to aid reformulation of the search terms, either expanding or restricting them.
- Use the definitions given in the index to the database.

8.3.1 CrossFire Beilstein*

The CrossFire Beilstein database derives from the Beilstein *Handbuch der Organischen Chemie* and, after *Chemical Abstracts*, is the most comprehensive compilation of organic chemical data. It is by far the best source for the chemical and physical properties of organic compounds: It contains about 9 million compounds, 9 million reactions and some 30 million other data. The main types of information held are structures and configurations; sources and isolation of natural products; synthesis, preparation and purification; characterisation and analysis; physical properties of pure compounds; reactions.

The data on these compounds derive from a critical selection from some 180 journals, as well as from patents (currently up to 1980) and monographs, and are fully referenced.

Beilstein is a suitable source for queries about synthetic procedures and physical properties of a compound, avoiding lengthy searches through the original literature, as its reliable information is often sufficient. Beilstein contains the most comprehensive

---

* Screens of CrossFire Commander, Version 7 for Windows are shown. For Mac OS, there is only a CrossFire Commander, Version 6 available.

collection of references to spectroscopic data (but not the actual spectra), to thermodynamic properties, and to natural products and is thus a significant information source not only for organic chemists but also for physical chemists, physicists, chemical engineers and pharmacologists.

How to use CrossFire Beilstein

(a) *CrossFire: Structure and Technical Features*
CrossFire is a client server database system. The database is provided to a central server for the university; depending on the individual licensing agreement, EcoPharm and Gmelin may be included. The end user needs a suitable connection to the network. CrossFire Commander software, which formats the search for the central server and prepares the results for retrieval, must be installed. Access is restricted by user IDs and passwords

Once the program has been started and the network connection established using the button Connect, CrossFire Commander provides the display below:

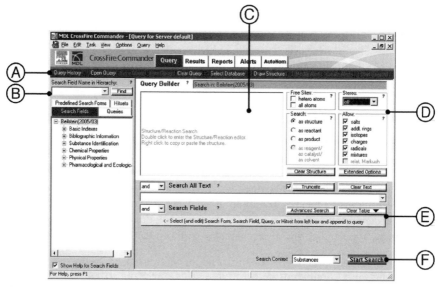

Figure 8.1: MDL CrossFire Commander. (CrossFire Software, 1995-2005, reproduced by permission of MDL Information Systems GmbH.)

Zones (A)–(E) contain the following functions:
(A) Buttons for loading and saving queries and for connecting to the server. Queries are accepted only after the user ID and password have been entered. The correct database (Beilstein or Gmelin) must also be selected using Select Database, to avoid searching the wrong one.
(B) Zone for selection of Search Fields, Queries, Predefined Search Forms, or Hitsets

(C) **Structure Editor** is used for drawing structures and is opened either via double-clicking the structure window (C), clicking button **Draw Structure** or using the menu **Task > Structure Editor**.

(D) Additional options for structure searches.

(E) **Fact Search Area**.

(F) **Start Search** initiates the search.

(b) *Types of Search*

CrossFire Beilstein is usually used to search for references to the synthesis, reactions and properties of a particular compounds. The example below uses TADDOL [1], a group of compounds often used as chiral auxiliaries, to illustrate some of the possible searches, and to show how progressive optimisation of the search terms can reduce the excessive number of hits.

(c) *Example of a Search*

(i) Search for the Synthesis and Properties of TADDOLs

Given: The structure of (R,R)-TADDOL (an abbreviation of (R,R)-α,α,α', α'-tetraaryl-2,2-dimethyl-1,3-dioxolan-4,5-dimethanol).

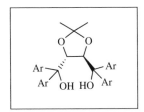

Figure 8.2: Structure of (R,R)-TADDOL

Search: Synthesis and properties of TADDOLs.

Strategy: Graphical input of the structure.

Inputting the structure: In **Commander**, open **Structure Editor** either using the button Draw Structure or via the menu **Task > Structure Editor** and draw the following structure as given in Figure 8.3

The structure is constructed using the following features from the horizontal list of functions and the vertical tool list:

(1) Icons for the elements C, O, N, S, P and a selection of other elements and generic groups (such as functional groups, alkyl residues and homo- or heterocyclic systems).

(2) Icons for single, double and triple bonds, together with a list of stereochemical indicators (**Up, Down**).

(3) Icon to quit **Structure Editor**.

(4) Ring template icons that place the selected ring on screen and activate the tool, which allows one to click in the middle of the ring in order to shift it to the desired location. The template is deselected by clicking on-screen.

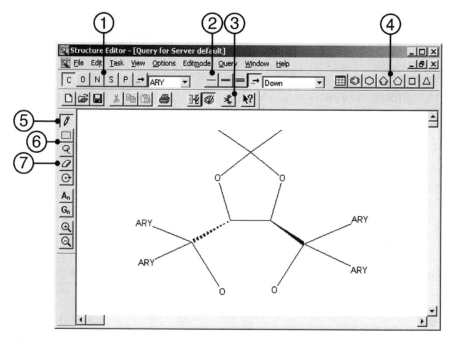

Figure 8.3: Structure Editor with the structural formula of TADDOL. (CrossFire Software, 1995-2005, reproduced by permission of MDL Information Systems GmbH.)

(5) A pen for drawing atoms, bonds and other structural features. Using this tool, the mouse arrow performs several functions:

    (a) It becomes an A when placed over an atom, a B when over a bond.

    (b) It confirms that a bond is being attached to a given atom by displaying the A.

    (c) Using the shift key and single-clicking it converts an element or bond into another selected from the list of functions.

(6) Rectangle or lasso icons can be used to select several atoms.

(7) The eraser icon is used to delete atoms or bonds.

The generic aryl groups are activated for use with the pencil as follows:

Double-clicking with the pencil on a terminal C atom (which the search will interpret as a methyl group) opens the **Atom Attributes** window (Fig. 8.4) for the selected atom. Choosing **Generics**...from the menu opens a catalogue of generic atoms and groups (Fig. 8.5) from which *ARY* (generic aryl group) is selected, and confirmed by *OK*.

In the **Atom Attributes** window, select **Set to Current** and confirm by *OK*. Shift clicking then converts all the selected terminal carbon atoms into aryl groups. All aryl groups, whether substituted or not, will then be included in subsequent searches.

Hydrogen atoms need not be drawn in using CrossFire Beilstein as they are automatically included. Once the structure has been drawn, one quits **Structure Editor** by

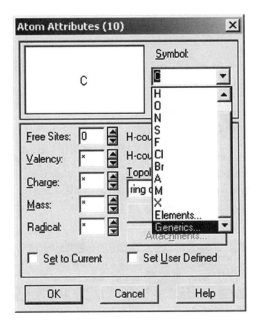

Figure 8.4: Atom Attributes window. (CrossFire Software, 1995-2005, reproduced by permission of MDL Information Systems GmbH.)

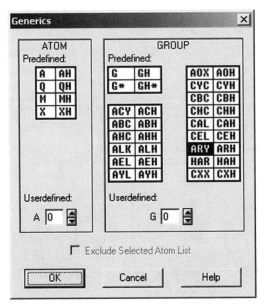

Figure 8.5: Generics window. (CrossFire Software, 1995-2005, reproduced by permission of MDL Information Systems GmbH.)

clicking the CrossFire icon; the structure is now exhibited in zone (C). In zone (D) the following parameters are set:

- Search: as *structure*
- Stereo: *off*
- Free sites: deactivate *hetero atoms* and *all atoms*
- Allow: tick all six boxes
- Search Context: *Substances*

Then **Start Search** to initiate the search. If successful, a new window, **Search Status Report**..., appears, displaying the input structure together with the number of hits. In this example, 149 hits were recorded [Update BS0503]. Clicking on the **View** button closes this window and opens a survey of the results in the **Results** window. **Short Display** exhibits the hits in blocks.

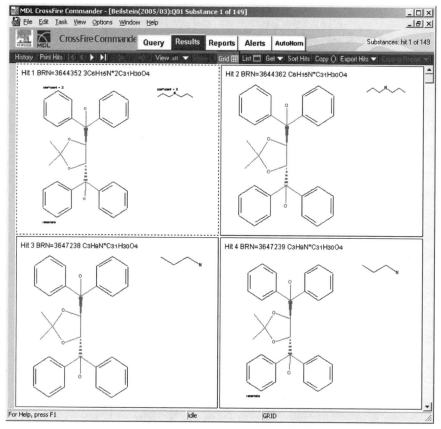

Figure 8.6: Display hits with a selection of the hits for TADDOLs. (CrossFire Software, 1995-2005, reproduced by permission of MDL Information Systems GmbH.)

Optimisation – Reducing the number of hits

On viewing the 149 hits, it is apparent that one has found not only the actual target but also TADDOLs in mixtures containing a variety of other materials as well as racemic forms. To exclude multicomponent systems, return to **Commander** and deactivate **Allow** in the six parameters in zone (D). This ensures that the next search will include only single compounds, excluding salts, isotopically labelled species, charged and radical derivatives and compounds with anellated aromatic rings.

This search generates 58 hits, and includes not only the input R,R structure but also R,S stereoisomers as well as compounds in which the stereochemistry is not defined, *i.e.* the program has ignored stereochemistry.

To set specific stereochemical requirements, the command **Stereo** = in zone (D) of **Commander** is set at *relative* instead of *off*. This gives 54 hits, selecting TADDOLs which are either R,R or S,S. Setting **Stereo** = *absolute* reduces this number to 41, as only the R,R isomers are then included.

Results Display

If one double-clicks on the formula of the first hit in **Grid Display** from the search with **Stereo** = *relative*, **List Display** opens and gives the following entry from the database.

```
Substance
Beilstein Registry Number      3657855
Beilstein Preferred RN         93379-48-7
CAS Registry Number            93222-42-5, 93379-48-7, 93379-49-8, 118139-82-5, 126639-52-9
Chemical Name                  (4R,5R)-bis(hydroxydiphenylmethyl)-2,2-dimethyl-1,3-dioxolane
Autoname                       [5-(hydroxy-diphenyl-methyl)-2,2-dimethyl-[1,3]dioxolan-4-yl]
                               diphenyl-methanol
Molecular Formula              C31H30O4
Molecular Weight               466.58
(....)
Reaction 1 of 75
Reaction ID                    1585226
Reactant BRN                   1098229 methanol
                               3657855 (4R,5R)-bis(hydroxydiphenylmethyl)-2,2-dimethyl-1,3-
                               dioxolane
Product BRN                    5844160 C31H30O4*CH4O
No. of Reaction Details        2
Reaction Classification
Other Conditions               Chemical behaviour
                               other alcohols, or amines, ketones, acetonitrile, nitromethane,
                               DMF, DMSO, THF, dioxane, benzene and other aromatics,
                               methylcyclohexane; other TADDOL's
Find similar reactions         not available
Ref. 1                         5708773; Journal; Weber, Edwin; Doerpinghaus, Norbert; Wimmer,
                               Claus; Stein, Zafra; Krupitsky, Helena; Goldberg, Israel; JOCEAH;
                               J. Org. Chem.; EN; 57; 25; 1992; 6825-6833.

Reaction 2 of 75
Reaction ID                    1604435
Reactant BRN                   3657855 (4R,5R)-bis(hydroxydiphenylmethyl)-2,2-dimethyl-1,3-
                               dioxolane
                               1098243 propylamine
Product BRN                    3647238 C31H30O4*C3H9N
```

Figure 8.7: Report of results for the TADDOLs (abbreviated). (Reproduced by permission of Beilstein Institut.)

Under Substance are listed data about the compound, including the Beilstein and CAS Registry Numbers, and its systematic name and Autoname (generated by the nomenclature program *Autonom*), as well as the date of its initial registration in the database and its most recent update.

Field Availability (accessed via the View > Field Availability Included menu if necessary) lists all the available data fields for the compound as well as the sections of the database containing relevant physical and chemical information. In order to avoid scrolling endlessly through all these fields, one can skip to the desired data as follows:

Choose menu View > Field Availability to view the window showing all the fields containing data on the desired compound in a separate window. The Occ(urrence) column lists the number of entries under each heading. For TADDOL this figure is 75 for

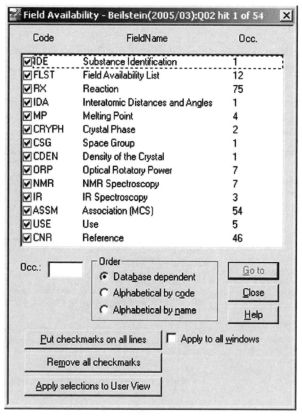

Figure 8.8: Field availability window. (CrossFire Software, 1995-2005, reproduced by permission of MDL Information Systems GmbH.)

entries dealing with reactions and 7 for NMR. Double-clicking on these rows displays the relevant data in the main window.

The Field Availability window can also be used to produce a printout of an individual User View. For example, to print out the entry under *NMR*, operate the Remove all checkmarks button, tick the relevant box in the properties field and then click on the Apply selections to User view button.
Caution: The selected User view will be saved and used for all later searches. To avoid confusion, cancel the selection after use via menu View > All Fields.

In the display of results (*e.g.* Reaction, Melting point) the underlined numbers are hyperlinks which can be clicked on to provide automatic connections to supplementary information in the database on particular compounds, reactions or references. The hyperlinks are colour-coded as follows:

- red: opens a Reaction Display window with details of the reaction.
- black: leads to a data set of an educt.
- blue: leads to a data set of a product.
- green: opens a Citation Display window containing bibliographical references and abstracts of articles in which the desired properties are described.

The list of functions includes the following additional useful tools:

○ Figure 8.9: Toolbar of functions in Results display. (CrossFire Software, 1995-2005, reproduced by permission of MDL Information Systems GmbH.)

(1) Navigates among the hits from an enquiry.
(2) Navigates within the window called up via a hyperlink.
(3) Displays the actual structure in a separate window.
(4) Switches between Grid View of structures and the List View of database entries for a structure.
(5) Copies the actual structure into Structure Editor.

Synthesis of TADDOLs

To select from the 54 hits those entries that describe syntheses of TADDOLs, one chooses from within Results Display the sequence View > Substance as Product only. This procedure reduces the number of hits under Field Availability from the original 75 to just 7.
Warning: This setting is saved.

(ii) Search by Name – a Comparison With the Structure Search

Search: TADDOLs using trivial names
Strategy: Predefined Search Forms > Substance Identification Data

In CrossFire Commander delete the structure used above using the Clear Structure button in zone (D). In tab Predefined Search Forms (zone [B]) double-click on Structure Identification Data to call up a form containing fields, *e.g.* Beilstein or Chemical Abstract Registry Numbers, chemical names and molecular formulae; enter 'taddol' in the Chemical Name input field. Alternatively, use the list hyperlink to call up the index of the Chemical Name field in the database. In the dialogue box Expand Chemical Name, type a few letters to jump to the correct section of the alphabetical index. Copy the name by double-clicking on the row containing the desired name and close the box by clicking *OK*. Confirming the entry in Substance Identification Data by clicking *OK* causes "Chemical Name (CN) is taddol" to appear in the table of queries in the Fact Search Area (zone [E]). The search can now be carried out.

The search results in only one hit, *viz.* the phenyl derivative, without any stereochemical indication in the formula. If one uses 'Chemical Name (CN) is *taddol** (*allows the presence of other symbols) as input instead, one eventually gets just 11 hits, including (+)-TADDOL, TADDOL and tetraphenyl TADDOL, clearly many fewer than the structure search. Therefore, a search using the name should only be used when one does not know the structure. Once a structure has been found it can be copied into Structure Editor using Copy Structure in Results Display, modified as desired and then used for a search using the structure.

(iii) Search for Reactions

Search: Reactions of TADDOLs with R-PCl$_2$ (for arbitrary R)
Strategy: Graphical input of starting materials and their reactions. Supplementary substructure search for R-PCl$_2$.

First clear the query table in zone (E) by clicking the button Clear Table > Delete All. In Structure Editor draw the structures of TADDOL and PCl$_2$ as shown below. Set Free Sites, arbitrary residues utilising the free valencies on phosphorus, by double-clicking with the pen on the phosphorus atom and then setting Free Sites: in the Atom Attributes window to *MAX*. To search for reactions, use icon (1) to select Reaction Edit in Structure Editor, or access it via the Editmode menu.

⭕ Figure 8.10: Standard Bar in Structure Editor. (CrossFire Software, 1995-2005, reproduced by permission of MDL Information Systems GmbH.)

(1) Icon for Reaction Edit mode
(2) Icon for Structure Edit mode
(3) Icon for CrossFire Commander

In Reaction Edit mode use the rectangle tool to select the TADDOL structure and define it as the starting material using the Reactant button. Repeat the procedure for the PCl₂ substructure. Then use the icon (3) to return to Commander.

Before starting the search, set the following parameters:

- Stereo: *relative*
- Free sites: deactivate *hetero atoms* and *all atoms*
- Allow: no ticks in any position
- Search Context: *Reactions*

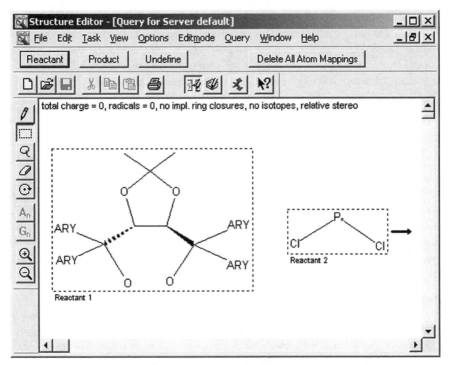

◯ Figure 8.11: Reaction Edit mode in Structure Editor. (CrossFire Software, 1995-2005, reproduced by permission of MDL Information Systems GmbH.)

This procedure gave 13 reactions including, e.g. the formation of a bicyclic phosphite by reaction with dichlorophosphate methyl ester [2].

(iv) Search for Properties

Search: TADDOLs whose NMR spectra have been measured

Strategy: Combined structure search and properties search.

The structure of TADDOL is set in Structure Editor as before. Return to Commander and choose the input mask Spectroscopic Data in Predefined Search Forms and tick

NMR before closing the window with *OK*. The query '*NMR spectroscopy (NMR) exists*' appears in the query table (zone [E]).

Set the following parameters:

- Stereo: *relative*
- Free sites: deactivate *hetero atoms* and *all atoms*
- Allow: no ticks in any position
- Search Context: *Substances*

This search gave 29 hits. To view only the structures and the NMR data, use the View > Hit only menu in Display Hits.
Warning: This setting will be saved for all subsequent searches until it is reset by View > All Fields. The display includes, *e.g.* the nucleus (^{1}H, ^{13}C, *etc.*), chemical shifts, coupling constants, solvent and temperature, but does not show the spectrum or individual measurements, although it does give references to the literature containing the spectra.

Further hints on CrossFire Beilstein searching can be found on the internet or via the Help function in the program.

8.3.2 CrossFire Gmelin

Gmelin Handbuch der Anorganischen Chemie, the inorganic chemistry equivalent of Beilstein, provides the source for the CrossFire Gmelin database. After *Chemical Abstracts*, it is the most comprehensive source of information about inorganic and metallo-organic compounds. It contains some 2.2 million substances, including coordination compounds, alloys, glasses, ceramics, minerals, isotopically labelled compounds, as well as 1.8 million reactions and 1.2 million references (November 2005).

A compound in CrossFire Gmelin can be restricted to the following categories:
- contains no carbon;
- contains carbon and at least one of the so-called 'Gmelin elements' (all elements in the Periodic Table except Li-Cs, Mg-Ba, Si, N-As, O-Te and F-I); the carbon-containing compounds also include modifications of elementary hydrocarbons, carbides, cyanides, and salts of 'simple' carboxylic acids, as well as small molecules such as H_2CO_3, CO_2, CS_2, *etc.*

CrossFire Gmelin is searched via the Commander software in a manner similar to CrossFire Beilstein; it comprises even more property fields. Because the printed *Handbuch* is not structured in a continuous manner, CrossFire Gmelin does not always contain composited information but often refers only to the printed version for many of the extensive records.

Example of a Search

Search: Use of TADDOLs as Catalysts
Strategy: Substructure search combined with property search

In Commander, select Gmelin in Select Database (zone [A]). In Structure Editor, draw the TADDOL structure as above, setting the number of free sites on the oxygen atoms of the terminal hydroxyl groups to *MAX* using the Atom Attributes window, and return to Commander. Select tab Search Fields (zone [B]), and find the desired property by entering 'catalyst' into the form below Search Field Name in Hierarchy and clicking the Find button. The field hierarchy is expanded and 'Behaviour as Catalyst' is shown as first hit in the tree. Single-clicking 'Behaviour as Catalyst' opens a window Field Help that displays information about this field. Double-clicking 'Behaviour as Catalyst' copies the field name to the table of queries in the Fact Search Area (zone [E]).

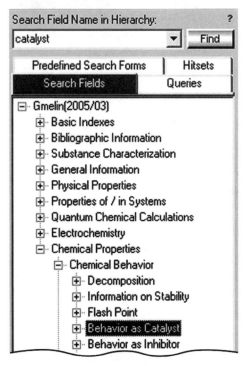

Figure 8.12: Hierarchical tree of database fields. (CrossFire Software, 1995-2005, reproduced by permission of MDL Information Systems GmbH.)

The principal buttons of the Fact Search area (Figure 8.13) are:

(1) Clear Table to delete table cells, rows or the whole table;
(2) bracket icons to group queries;
(3) the expand list icon for displaying and selecting values of a chosen property.

The following search options are then set:

- Stereo: relative
- Free sites: no ticks

- Allow: tick all boxes
                         This will also generate structure searches for multicomponent mixtures, salts and annelated rings; it is significant that one may not always know exactly what complex compounds are included in Gmelin.
- Search Context: Substances

○ Figure 8.13: Fact Search Area. (CrossFire Software, 1995-2005, reproduced by permission of MDL Information Systems GmbH.)

This search gave 29 hits (Gmelin Update GM0503), including 27 Ti(IV) organic complexes with TADDOL as a chelating ligand and two mixtures containing Mo^{6+}, TADDOL and other components.

8.3.3 SciFinder Scholar

SciFinder Scholar™, provided by CAS, uses efficient, user-friendly software to search *Chemical Abstracts*, the most comprehensive review of the chemical literature. *Chemical Abstracts* (CA) comprises related databases of references, structures and reactions, in which publications can be searched by keywords, authors, compounds (including polymers, industrial materials and biopolymers such as peptides and nucleic acids) reactions and (from 1997) citations. The database is set up by continuous evaluation of some 9000 journals, which is why it is vital for a comprehensive literature search.

SciFinder Scholar provides access to the following databases:

- CAplus^SM: a database of more than 25 million references from journal articles and patents since 1907. The abstracts are taken directly from the publications and contain author, title and source as well as a summary of the contents. In addition, the papers are correlated through some of the 80 CA sections (e.g. Section 23, Aliphatic Compounds), keywords and compound classes (from a thesaurus), Registry Numbers of compounds and (since 1998) by citations in the paper.
- CAS REGISTRY^SM: a database of some 26 million compounds and 56 million biological sequences, giving their Registry Number, chemical formula, name and structure. Since 1967, each definable chemical compound has been assigned a CAS Registry Number: stereoisomers, racemates, compounds with undefined stereochemistry, salts, ions, mixtures, radicals and isotopomers each have their own Registry Numbers; sometime a single compound may have more than one Registry Number. The Registry Number is often used in other databases and is therefore a useful search term for locating a compound rapidly.

- CASREACT®: a database containing some 10 million single or multistage reactions. The bonds that are formally broken or formed in starting materials and products are identified. A multistep reaction, A → B → C → D, is indexed in such a way that in addition to searches for A → D, searches for intermediate steps such as A → C or B → D are possible. In addition, the reaction type and the yield are indicated.
- CHEMCATS: this is a catalogue of about 9 million chemicals, giving supplier information.
- CHEMLIST: references to legal definitions concerning chemicals.
- MEDLINE: the database of the US National Library of medicine, containing reports of about 15 million medical and biological publications [A free version, PubMed, is available at www.pubmed.org].

SciFinder Scholar is structured in such a manner that a search is actively promoted stepwise along a particular pathway; at times one can return to the previous step along this pathway and adjust the enquiry. The results from different pathways cannot be combined, which makes it different from Beilstein and other databases. It is thus advisable to begin as broadly as possible. In each individual step, choices and refinements are continuously offered, so that one can use these to restrict the outcome. SciFinder also has analytical functions that can determine the frequency of, *e.g.* the names of authors or types of document. Using these, one can generate criteria for further reductions in the number of references.

How to use SciFinder Scholar

(a) *Technical Features*

SciFinder Scholar is the software for the CA client-server database system, which operates from a CAS central server in Columbus, Ohio, where the searches are processed. To log on one needs to have the licensed software installed and must have a network address at the university holding the licence.

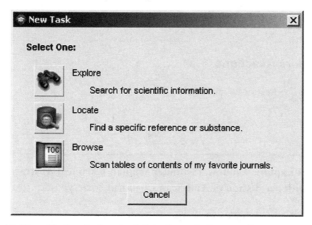

Figure 8.14: New Task window in SciFinder Scholar. (Scifinder Scholar, Chemical Abstracts Service, reprinted with permission of American Chemical Society)

On logging on, the licence conditions must first be ratified with the server in Columbus. A **Message of the Day** window appears containing notices from CAS. Click *OK* to bring up the **New Task** window which offers three different tasks, **Explore, Locate** and **Browse**. The latter should not be used because one of the few concurrent-user licences will be blocked. Instead of this option, the local library list of e-journals can be used.

(b) *Example of a Search*

(i) Search for a compound

Search: papers on the influence of TADDOL aryl substituents on the stereoselectivity of reactions

Strategy: substructure search followed by restriction of the results using a search for themes. The substructure search is done in CAS Registry, the theme search in CAplus.

Choose **Explore** in the **New Task** window to raise a secondary window, **Explore**. Experience suggests that the functions **Chemical Structure** or **Reaction Structure** are the most profitable; **Research Topic** provides a search for themes. The **Chemical Structure** icon opens the SciFinder Scholar structure editor (Figure 8.16).

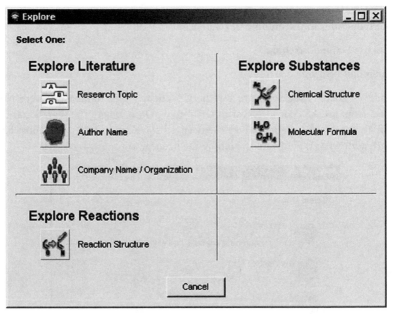

○ Figure 8.15: Explore Window. (Scifinder Scholar, Chemical Abstracts Service, reprinted with permission of American Chemical Society)

On the left-hand side of this window are the tools for drawing structures. Icons for the most common elements and types of bond, as well as for starting the search are found at the bottom. If the mouse arrow is placed on a button, after a short pause a yellow *tool tip* appears, giving a short description of its function.

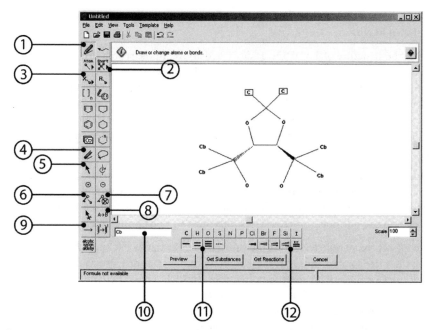

○ Figure 8.16: Window for entering structures in SciFinder Scholar. (Scifinder Scholar, Chemical Abstracts Service, reprinted with permission of American Chemical Society)

The main tools as shown in Figure 8.16 are:

(1) pen
(2) shortcut to predefined functional groups
(3) icons for variable groups: X = optional halogen; A = element other than H; Q = element other than H or C; M = optional metal; Ak = alkyl chain; Cy = ring structure; Cb = carbocycle; Hy = heterocycle
(4) eraser for deleting atoms or bonds
(5) icon for moving individual atoms or bonds
(6) Atom Locking Tool, which excludes particular atoms from a substructure search: atoms blocked from substitution are shown in a box
(7) Ring Locking Tool, which excludes rings from a substructure search (preventing the formation of annellated rings): blocked rings and chains are shown in bold
(8) reaction role tool, with which a compound can be defined as, *e.g.* reactant, reagent or product: this involves clicking on the icon and then on the compound, choosing the function from the dialog box which appears
(9) arrow for indicating the direction of the reaction, thus defining starting materials and products
(10) indicates which element or group is activated by the pen
(11) icons for types of bonds
(12) icons for specifying stereochemistry

When the mouse arrow is placed over a bond, atom, ring or chain, the chosen object is shown in red.

Draw the structure of TADDOL as described above. In the search this will be input as a partial structure, *i.e.* with free valencies on all atoms. Atoms that are to be excluded from the substructure search must be blocked using the **Atom Locking Tool** (6); this is the opposite of the CrossFire procedure where available **free sites** have to be set directly. In this example, we only want the aryl groups to be variable, so the methyl groups need to be surrounded by a box to avoid searching for other alkyl groups.

The search is initiated using the **Get Substances** button, which opens a window in which one can specify an exact match (with H filling all free valencies) or a substructure search; as the input structure already shows variable groups, a substructure search is already activated. Both mixtures and compounds will be included in the search; to exclude one of these classes, this must be done using **Filters**.

Having set the parameters, start the substructure search by clicking *OK*. It is possible that there are so many hits that the system limits are exceeded, especially with smaller molecules or those with too many free sites. In such a case, SciFinder Scholar sends an error message saying that the input structure is too general and suggests restricting further additional atoms or excluding annelation (**AutoFix** prompts options) or modifying the search either using **Additional Options** or altering the molar mass.

In this search (November 2005) 862 structures were found in the CAS Registry, sorted by stereochemistry (Figure 8.17). As all these hits might be relevant in answering the

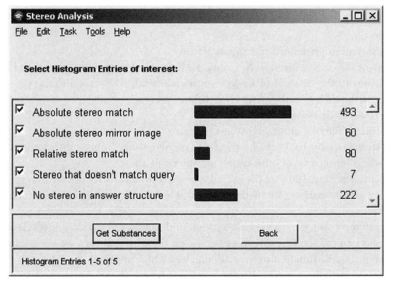

Figure 8.17: Stereo Analysis dialog box in SciFinder Scholar. (Scifinder Scholar, Chemical Abstracts Service, reprinted with permission of American Chemical Society)

question of the influence of the aryl substituents on stereoselectivity, all boxes were ticked and the relevant structures called up using the **Get Substances** button.

The formula overview gives the CAS Registry Number and number of references for each compound in a block, with the partial structure highlighted in red.

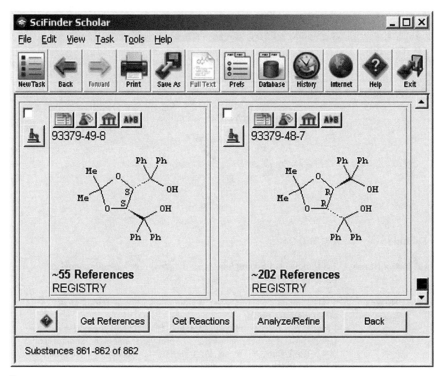

Figure 8.18: Window displaying the structure hits found. (Scifinder Scholar, Chemical Abstracts Service, reprinted with permission of American Chemical Society)

The CAplus^{SM}, CHEMCATS®, CHEMLIST® and CASREACT® databases contain entries for each of these compounds which can be called up using the relevant icons:

All abstracts containing the particular compound (CAplus^{SM})

Reference information and manufacturers (CHEMCATS®)

National and international regulations (CHEMLIST®)

Reactions in which the compound is present (CASREACT®). In this case, the **Reaction Roles** window, in which the role of the compounds (whether starting material, product *etc.*) must be defined, opens.

Clicking on the *microscope* icon opens a window containing the CAS Registry^{SM} entry. For (*R,R*)-tetraphenyl-TADDOL this gives the CAS Registry Number, the structural formula and the CA Index name as well as others including trivial and trade names. There follow hyperlinks for lists of calculated characteristics (e.g. bioconcentration factor, boiling point, enthalpy of vaporisation, molar solubilities, pK_a) as well as measured properties (melting point, optical rotation). At the end of the entry is a list of databases containing information about TADDOL.

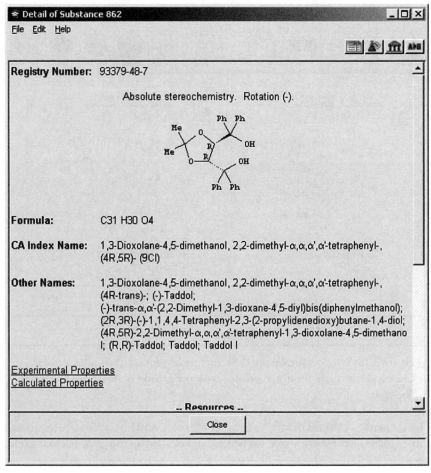

Figure 8.19: CAS Registry[SM] entry for TADDOL. (Scifinder Scholar, Chemical Abstracts Service, reprinted with permission of American Chemical Society)

Within the formula overview there are links to both Get References and Analyze or Refine Substances. The latter leads to additional dialog boxes in which the identified compounds can be limited according to various structural features (substituents, tautomerism [Analyze by Precision], ring systems or stereoisomerism) by partial structure, commercial availability or physical characteristics. It is worth exploring these options at least once.

Using **Get References** the abstracts relating to the chosen compounds are displayed. The references displayed in the **Get References** window can be limited in advance (*e.g.* to manufacturers). The suggested settings (all substances, all references) are confirmed by *OK*. The search in CAplus gave 435 references (November 2005) which mention the 862 compounds noted above. The bibliographical data are listed in descending order of dates (but see below).

These references are not yet ready for examination as the next step is to limit the list to those answering the question of the influence of the aryl substituents on stereoselectivity. For this, one chooses the **Analyze/Refine** button at the bottom of the window followed by **Refine** in the new **Analyze or Refine** window. In the **Refine** window there are a number of options for limiting the number of references, *e.g.* themes, specific authors, date of publication or type of document.

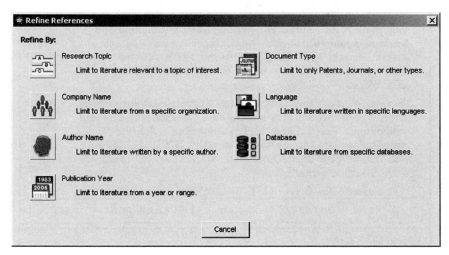

Figure 8.20: Refine References dialog box. (Scifinder Scholar, Chemical Abstracts Service, reprinted with permission of American Chemical Society)

Clicking the **Research Topic** button opens the dialog box **Refine by Research Topic** in which *substituent effect on stereoselectivity* is entered in the form and confirmed by *OK* (for searches for themes, see below). This generates 18 hits including a US patent [for free access to patents see http://espacenet.com] as well as journal articles. The hits include publications which contain variations of the CH_3 groups in the 2-position, as they include those in which TADDOLs with any substituent are listed, including 2,2-dimethyl-TADDOLs, and thus satisfy the search terms.

Using the microscope icon, the detailed entry for each publication can be called up in CAplus. The button showing a document takes one to **ChemPort Service** which can

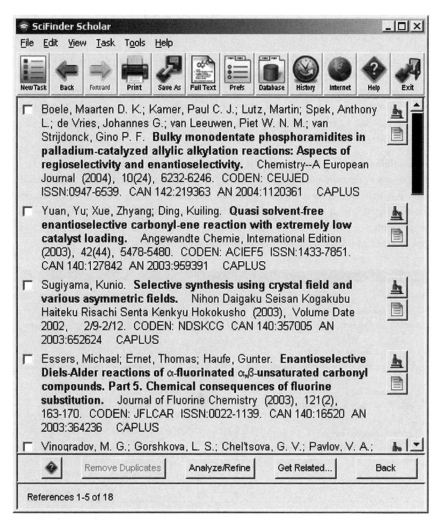

connect the user directly to an article in an electronic journal, a patent, a catalogue entry in a library or a costed delivery service.

(ii) Search for a Reaction

Search: Reactions of TADDOLs with R-PCl$_2$ (for arbitrary R)

Strategy: Graphical input of starting materials and their reactions. Supplementary substructure search for R-PCl$_2$. The search takes place in CASREACT(R), with output of relevant publications in CAplus^{SM}.

As before, call up the SciFinder **Structure** and draw the reaction shown in Figure 8.22.

○ Figure 8.22: Graphic representation for the reaction. (Scifinder Scholar, Chemical Abstracts Service, reprinted with permission of American Chemical Society)

By inserting the arrow (button 9) both compounds are automatically defined as re-actant/reagent. However, SciFinder Scholar interprets these roles rather freely unless they are restricted, so that it would, *e.g.* allow a starting material containing only one Cl atom, so one uses the reaction icon (8) to set PCl_2 as a reactant. Use the Get Reactions button to initiate the search in CASREACT®, confirming the substructure that appears in the Get Reactions window by *OK*.

SciFinder warns that (unlike Beilstein) no attention is paid to the indicated stereo-chemistry during the search. It follows the overview of the reactions that have been found by indicating that additional information about each compound (entry in CAS REGISTRY^{SM}, references, preparations and consecutive reactions) can be obtained.

This search gave ten hits, displayed in the overview in date order of their first appear-ance in the primary literature. Each reaction carries a hyperlink to the original refer-ence, which can be used to call up the relevant entry in CAplus^{SM} (Figure 8.23).

Using the Refine Reactions button the results can be further limited or refined, e.g. by selecting partial structural characteristics, yields, reaction type or whether a single or multistep procedure has been followed. The Get References button leads to four references for the ten reactions identified.

(iii) Search for a Theme

Search: The names of the most important authors working on enantioselective synthesis using TADDOLs

Strategy: Theme search via synonyms, analysing the results by the frequency with which authors are cited. The search takes place in CAplus^{SM}.

Click on Research Topic in the Explore window and enter a question in the Explore by Research Topic window which appears, using ordinary English to express the query in either extended or abbreviated form. Start the search with *OK*.

Using artificial intelligence, SciFinder Scholar® analyses the terms in the input, con-siders possible variants and alternatives and organises the information into so-called 'concepts'. It then displays a register of hits containing various combinations of these

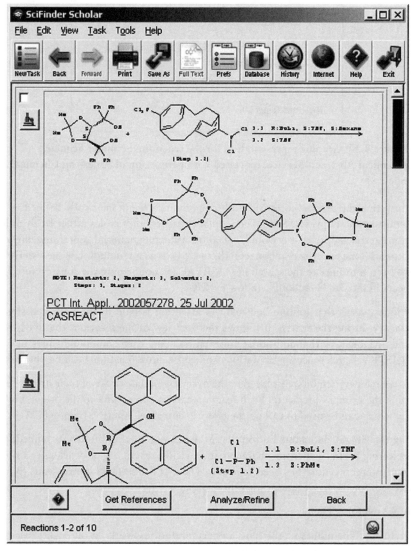

○ Figure 8.23: List (abbreviated) of reactions found. (Scifinder Scholar, Chemical Abstracts Service, reprinted with permission of American Chemical Society)

concepts. Be warned that artificial intelligence frequently gives misleading results, so the output should always be evaluated carefully. It is preferable to vary the input query to obtain the most appropriate results. Explore by Research Topic searches by title, abstract and in the index of not only CAplus^{SM} but also MEDLINE.

The following table shows how SciFinder Scholar® interprets various search inputs and gives hints on how to formulate an effective query.

Feature	Explanation and examples
Concept	Concepts are identified from their connecting prepositions and conjunctions. It is not advisable to include more than four concepts as the number of possible combinations would become too large. If two or three words follow one another without being separated by prepositions, they are treated as a single concept, but if there are more than three such words, they are separated. Example: substituent *effect on stereoselectivity* is interpreted as concepts *substituent effect* and *stereoselectivity* (and abbreviated forms). One can also input compounds as their CAS Registry Numbers.
Truncation	SciFinder Scholar will frequently identify the word at the root of a concept automatically together with its derivatives. These variants are treated as the same concept. However, the truncation algorithm is imperfect, so results need to be scanned carefully for errors; synonyms may be useful in expanding the search. Example: *catalysis* also finds *catalyst, catalytic* and *catalysed.*
Synonyms	SciFinder Scholar automatically finds synonyms and abbreviations. To some extent, additional synonyms may be given (in brackets, separated by commas), up to a maximum of four. Example: *enantioselective (asymmetric, stereoselective).* SciFinder Scholar associates an adjective only with the first synonym, so it has to be repeated. Example: input *catalytic reduction (catalytic hydrogenation),* not *catalytic reduction (hydrogenation)*
Boolean Operators	The Boolean NOT is the only operator consistently interpreted: NOT (placed *before* a concept). Excludes concept from consideration. Example: *taddol NOT taddolates*
Stop words	SciFinder Scholar has an internal list of words that are not considered in the search. Examples: *'information on', 'effect of.'*

Using these rules, the search question can be framed as follows: *taddol in enantioselective (asymmetric) catalysis.* SciFinder analyses this into three concepts, *viz. taddol, enantioselective/asymmetric* and *catalysis.* These are linked to generate 25 different hits, which are displayed as a list with the appropriate references. The hits are ordered by the number of linked concepts, weighted by the closeness of their association. Hits that contain only one of the concepts but have many references are presented at the end.

The first four hits are illustrated in Figure 8.24.

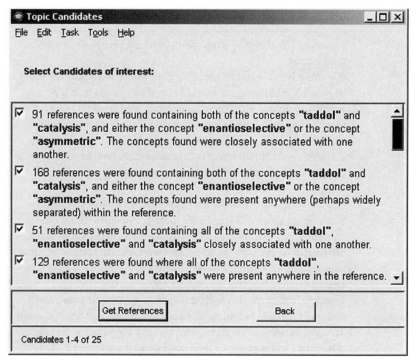

Figure 8.24: The first 4 or the 25 hits for the Topics chosen. (Scifinder Scholar, Chemical Abstracts Service, reprinted with permission of American Chemical Society)

The following linkages are commonly found.

Link	Meaning
as entered	The phrase occurs exactly as input in either the title, abstract or index section.
closely associated with one another	The listed concepts are found in every case either in the title or abstract, or in the corresponding section of the Index.
present anywhere in the reference	The listed concepts are divided between the title, abstract and Index. The supplementary phrase *perhaps widely separated* indicates that there may not necessarily be a connection between the concepts.
were found containing the concept XY	The concept XY occurs somewhere in the title, abstract or index.
but not	This is associated with the Boolean NOT: Contains concept A but not concept B.

The first 14 candidates which contain taddol and at least one other concept are marked with a tick in the summary window. The Get References button then produces the 243 literature references already displayed in the summary. The search terms that have been found are highlighted in blue in this summary, as they also are in the detailed database entry (accessed via the 'microscope' icon). After removing the duplicate references that occur in both CAplus^{SM} and MEDLINE with Remove Duplicates, one obtains 232 literature references.

These references can be investigated in more detail by clicking on Analyze in the Analyze or Refine References dialog box to reveal the Analyze window, which offers a variety of modes of analysis. It is particularly useful to examine references using, *e.g.* CA Section Title or Index Term, as these may indicate themes or keywords which can be used in subsequent searches.

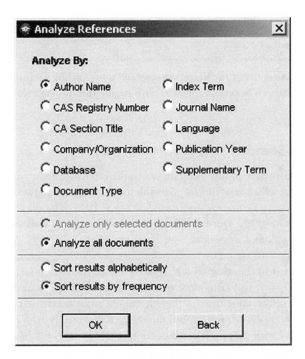

Figure 8.25: Analyze dialog box. (Scifinder Scholar, Chemical Abstracts Service, reprinted with permission of American Chemical Society)

Clicking OK on the default, Author Name, produces a histogram showing the number of references for each individual author. The first five of the relevant 612 authors are ticked; the Get References button then causes a total of 67 references by the five most significant authors to be displayed.

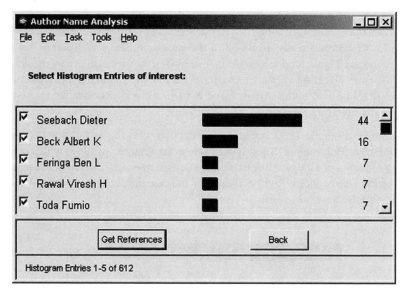

○ Figure 8.26: Analysis of the literature references by Author. (Scifinder Scholar, Chemical Abstracts Service, reprinted with permission of American Chemical Society)

8.3.4 Web of Science

Web of Science provides an internet gateway to the interdisciplinary *Science Citation Index Expanded*, which contains information drawn from the natural and biological sciences, technology, agronomy and medicine since 1945. The database consists of citations from some 5700 journals, giving a list of sources as well as the references they cite. Abstracts and keywords have been included since 1991. Web of Science can be used to investigate many interdisciplinary issues via author or title searches. It can also be used for citation searching, which is particularly useful if one wishes to use a seminal paper to locate more recent, related publications and to identify active researchers in the same field. Although the chemical literature only contributes about 15% of the material, substantial numbers of chemically relevant publications are cited, together with the references they contain (*Note*: Only the *International Edition* of *Angewandte Chemie* is included).

How to use Web of Science?

(a) *Structure of Science Citation Index*

Logging of references: These include names of all authors, their institutions, full journal title, year of publication, volume and section, as well as the first and last page numbers.

Logging of citations: These consist of only the first author and his/her institution, an abbreviated journal title, year of publication, volume, section and first page number only.

Treatment of authors' names:

- Only the initials of forenames are retained: Barton, Derek Harold Richard → BARTON DHR.
- Spaces in multipart names are deleted: de Man → DEMAN.
- Apostrophes and hyphens are deleted and non-ASCII symbols are transliterated: Müller-Herold → MULLERHEROLD.
- Names are truncated after 15 characters.

Problems with Citation Searching
Because authors names are used in abbreviated form and references are not unequivocally assigned with the desired author placed first, there are often problems with citation searching.

Other problems: Only journals are included in the database, not monographs. References are frequently cited incorrectly in the literature. There is also the fundamental fact that, as shown by statistical surveys, within the first 5 years after publication, papers are cited only once in over half the relevant publications and not at all in about a fifth of them.

(b) *Example of a search*

Organic chemistry textbooks refer to the Sequence Rules for specifying configurations laid down by Cahn (R.S.), Ingold (C.K.) and Prelog (V.). Web of Science can answer questions such as the following:

- Is Cahn, R.S. still being cited in the primary literature? If so, in which contexts?
- Can one locate the original Cahn-Ingold-Prelog paper(s)?

To answer the first of these questions, load the Web of Science website via your internet browser (a license is required). Use the Full Search button to obtain the Search box and choose CITED REF SEARCH. In the resulting window enter *CAHN RS* as the CITED AUTHOR and click on *LOOKUP* to start the search. The Cited Reference Search page displays the first 20 hits out of 197 papers in which *CAHN RS* is cited. Among these, and the first 20 hits out of 207 references are underlined, meaning that there is no hyperlink from them to a Cahn paper. This indicates that either the relevant paper by Cahn has been incorrectly cited, or else the citation is to a paper in a publication that is not included in the Web of Science database. Using the forward arrow to display the next 20 hits, one obtains several that have hyperlinks.

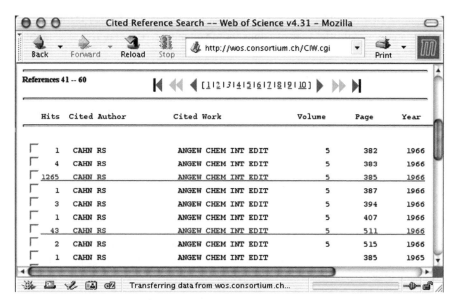

Figure 8.27: List of hits no. 41-60 for the citations found. Information was taken from Thomson Web of Science. (Reproduced by permission of Thomson Scientific)

Clicking the row which shows 1376 citations (as of March 2, 2007) leads to a detailed data entry for the desired paper. As it was published before 1991, neither an abstract nor keywords are provided.

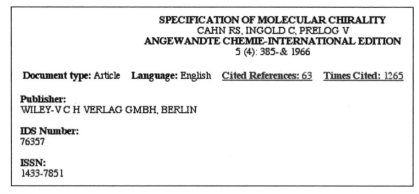

Figure 8.28: Window showing the details of the publication. Information was taken from Thomson Web of Science. (Reproduced by permission of Thomson Scientific)

Using the *Cited References* hyperlink one obtains the 63 references quoted in this article; the *Times Cited* link gives all 1376 papers in which the article has been cited to date.

If one uses *PRELOG V* as the CITED AUTHOR instead one obtains 927 references, of which the window showing numbers 81–100 provides a hyperlink to the same article.

To answer the second query, one selects GENERAL SEARCH on the Full Search page. Clicking on the hyperlinks Topic, Author, Source and Address in the GENERAL SEARCH window leads to 'Help' pages with information on the relevant search possibilities. Typing *Cahn RS and Ingold CK and Prelog V* in the input line Author and clicking on SEARCH initiates the search, which leads to a single hit, the desired original paper: *The Specification of Asymmetric Configuration in Organic Chemistry, Cahn RS, Ingold CK, Prelog V, Experientia 12 (1956), 81–94.*

8.4 REVIEWS

8.4.1 Annual Reports in Organic Synthesis (Academic Press)

In this annual series, founded in 1970 (most recent volume 2004), useful synthetic methods taken from the primary literature are arranged under headings such as C–C bond formation, oxidation or reduction. Synthesis of particular classes of compounds such as heterocycles, use of protecting groups and review articles are included. Information on recent methods of synthesis can be derived rapidly from the graphical schemes, which consist of the structures of starting materials and products, with the reagents and reaction conditions indicated over the arrow.

8.4.2 Compendium of Organic Synthetic Methods (founded by I. T. Harrison and S. Harrison, Wiley)

There are currently 11 volumes of this compilation of 'functional group interchanges' and methods for preparing mono-and difunctional compounds, covering the most important functional groups. The entries comprise graphical schemes and literature references, based on work published in the previous three years (Vol. 11 appeared in 2003).

8.4.3 Comprehensive Organic Transformation: A Guide to Functional Group Preparations (R.C. Larock, 2nd edition, Wiley, 1999)

Similar to *Compendium of Organic Synthetic Methods*, arranged by reaction type.

8.4.4 Organic Reactions (founded by R. Adams 1942 (Vol. 1; 2005: Vol. 82))

In each volume particular types of reactions that are of general application are discussed with emphasis on preparative aspects, with references to the original literature. Individual chapters contain comprehensive tabulated compilations.

8.4.5 Organic Syntheses (founded by R. Adams 1921 (Vol. 1; 2005: Vol. 81))

Volumes appear annually and contain tested "recipes" for syntheses of general interest. Ten collective volumes have also appeared, each containing the corrected material from the annual volumes.

8.4.6 Reagents for Organic Synthesis (founded by M. Fieser and L. F. Fieser)

There are currently (2004) 22 volumes of this series, comprising an inventory of reagents from the recent literature, with references and examples of applications. A collective index for Vols. 1–22 was published in 2005.

8.4.7 Comprehensive Organic Functional Group Transformations II (Vols. 1–7, A. R. Katritzky, R. J. K. Taylor, Elsevier, 2004)

An entry to the literature of functional group transformations. It provides information about the organic synthesis and interconversion of all known functional groups.

8.4.8 Comprehensive Organic Synthesis (Vols. 1–9, B. Trost and I. Fleming, eds., Pergamon, Oxford, 1989)

An overview of all the most significant synthetic methods and reagents, with many examples of their application. Arranged by reaction type (addition to $C=C$ and $C=X$, C–C bond formation, introduction and interconversions of functional groups, oxidation, reduction). The individual chapters by recognised experts provide detailed information on the scope of the particular reagents or method.

8.4.9 Comprehensive Organic Chemistry (Vols. 1–6, D. H. R. Barton and W. D. Ollis, eds., Pergamon, 1979)

This comprises the synthesis and reactions of organic compounds.

8.4.10 Comprehensive Organometallic Chemistry II (Vols. 1–14, E. W. Abel, F. G. A. Stone, G. Wilkinson, eds., Pergamon, 1995); Comprehensive Organometallic Chemistry III (Vols. 1–13, R. Crabtree, M. Mingos, eds., Elsevier, 2006)

Discusses the synthesis, reactions and structures of organometallic compounds.

8.4.11 Science of Synthesis, Houben-Weyl Methods of Molecular Transformations, Methoden der Organischen Chemie

'Houben-Weyl' contains tested synthetic methods for the most important classes of compounds. There are some 50 volumes of this comprehensive work, covering general, analytical and physical techniques as well as oxidation, reduction and photochemical reactions. A web version is available.

8.4.12 Synthetic Methods for Organic Chemistry (founded by W. Theilheimer)

The annual supplements (2005: Vol. 67) contain short descriptions of selected reactions from the recent literature together with experimental details. 'Theilheimer' is effective for locating the original references to a given type of reaction. Each volume contains an index of compound types; reagents are entered separately. The classification system divides reactions into four types, using a cipher for each.

Reaction Type	Cipher
Addition	⇑
Rearrangement	⋔
Exchange	↑↓
Elimination	⇓

The symbols of the atoms between which a new bond is formed in the reaction product are placed before each cipher, the affected bonding in the reactants being shown after it.

The system of classification is explained in Vol. 2; once grasped, the systematic surveys and supplementary references become useful. There is a cumulated index in every fifth volume.

8.4.13 Techniques of Chemistry (A. Weissberger)

Fourteen volumes covering physico-chemical methods, separation, analytical techniques, reaction mechanisms and organic photochemistry. Revised second or third editions of individual volumes appeared prior to 1980.

8.4.14 Specialist Periodical Reports of the Royal Society of Chemistry (The Chemical Society, London)

Given the increasing number of special areas of interest and the flood of publications, these annual or biennial reviews of a large variety of fields constitute a reliable source of information. Topics covered include amino acids, carbohydrate chemistry, catalysis, electron spin and nuclear magnetic resonance spectroscopies, organometallic and organophosphorus chemistry and photochemistry. Volumes published from 1998 onwards are available electronically (a licence is required).

8.5 MONOGRAPHS, REFERENCE WORKS AND HANDBOOKS

There are so many monographs covering such a vast range of subjects, it would be best to consult a specialist (or fellow student!) for specific purposes. However, searching the catalogues of local institute libraries, either alphabetically or in the keyword index (using names of authors, publishers or titles as input) may be effective.

Every library in a chemistry department has reference works for classes of compounds, synthetic methods, laboratory techniques and separation methods as well as information about the toxicological properties of chemicals. Particularly noteworthy is Bretherick's Handbook of Reactive Chemical Hazards, P. Urben and M. J. Pitt, 7th edition, Elsevier Academic Press, 2007. There are many useful handbooks commonly used in laboratories. These include:

8.5.1 CRC Handbook of Chemistry and Physics (often called the 'Rubber Bible').

Issued biennially (87th edition, 2006), this work tabulates mathematical formulae, chemical elements and their isotopes and properties and physical constants, as well as the physical properties of organic, inorganic and metallo-organic compounds.

8.5.2 Merck Index (14th edition, 2006)

The Index is effectively an encyclopaedia of chemical, pharmaceutical and biological products, listing some 10 000 entries, including synonyms, references, information on toxicity and therapeutic applications. There are also sections giving the CAS Registry Numbers of the relevant compound, a catalogue of named reactions and a list of cooling mixtures. It is available as a CD or on-line.

8.5.3 Heilbron's Dictionary of Organic Compounds

This comprehensive reference work lists a multitude of organic compounds alphabetically by name, together with their structural formulae. In addition, it also gives melting and boiling points, optical rotations where relevant, stability data, CAS Registry Numbers and references to recent literature. The 6th edition (1996) comprised 10 volumes; there have been two further supplements.

8.6 DATA COLLECTIONS

The fact that there are so many journals that contain spectroscopic and physico-chemical data in original papers increases considerably the difficulty of locating such information on a particular compound or class of compounds. The identification of organic compounds is becoming increasingly dependent on instrumental methods of analysis. The use of these methods is substantially aided by monographs and collections of data, especially those which are updated regularly.

The following list indicates some of the current data collections on NMR, IR, UV/visible and mass spectroscopy.

NMR Spectroscopy

- The Aldrich Library of ^{13}C and ^{1}H FT-NMR Spectra, Vols. 1–3, C.J. Pouchert and J. Behnke, Aldrich Chemical Co., Milwaukee, 1993.
- The Aldrich Library of NMR Spectra, 2nd edition, C.J. Pouchert, Aldrich Chemical Co., Milwaukee, 1983.

- Sadtler Standard Nuclear Magnetic Resonance Spectra, 25 Vols. (from 1966), Sadtler Research Laboratories, Philadelphia.
- Carbon-13 NMR Spectra: A Collection of Assigned Coded and Indexed Spectra, L.F. Johnson and W.C. Janukowski, Kriegel Publishing Co., New York, 1978.
- Atlas of Carbon-13 Data, Vols. I–III, E. Breitmeyer, G. Haas, and W Voelte, Heyden, London, 1979.
- Carbon-13 NMR Spectral Data, 4th edition, W. Bremser, L. Ernst, W. Fadinger, R. Gerhards, A. Hardt, and P.M.E. Lewis, VCH-Verlag, Weinheim, 1987 (contains some 60 000 spectra).

IR Spectroscopy

- The Aldrich Library of FT-IR Spectra, 2nd edition, Aldrich Chemical Co., Milwaukee, 1997.
- The Aldrich Library of Infrared Spectra, 3rd edition, C.J. Pouchert, Aldrich Chemical Co., Milwaukee, 1981.
- Sadtler Standard Spectra (from 1962), Sadtler Research Laboratories, Philadelphia.
- DMS Atlas of IR Spectra, Verlag Chemie, Weinheim, 1972.

Mass Spectrometry

- Eight Peak Index of Mass Spectra, 3rd edition, 8 Vols., Royal Society of Chemistry, Cambridge, 1991.
- The Wiley/NBS Registry of Mass Spectral Data, F.W. McLafferty, D.B. Stauffer, 7 Vols., Wiley, New York, 1989.
- NIST/EPA/NIH Mass Spectral Library with Search Program, 2005, National Institute of Standards and Technology, US Department of Commerce.

UV/Visible Spectroscopy

- Selected UV Spectral Data (1945–1984; 1985–1999), 10 Vols., Thermodynamics Research Center.
- Organic Electronic Spectral Data (Vols. 1–31, 1946–1989), H.P. Phillips, H. Feuer, and B.S. Thyarajan, Wiley-Interscience, New York.
- UV-Vis Atlas fo Organic Compounds, H.-H. Perkampus, Wiley-VCH, 1992.

Web Sources for Chemical Spectra and Spectral Data

NMR
http://www.nmrshiftdb.org/

NMRShiftDB is a web database for organic structures and their nuclear magnetic resonance (NMR) spectra. It allows for spectrum prediction (currently only for carbon-13) as well as for searching spectra, structures and other properties. In addition, it features peer-reviewed submission of datasets by its users. The NMRShiftDB software is open source; the data are published under an open content licence.

IR, MS and UV/Vis spectra
http://webbook.nist.gov/

The National Instiute of Standards and Technology provides free access to the NIST Chemistry WebBook, which contains information on IR ($>16\,000$ compounds), MS ($>15\,000$ compounds), UV/Vis spectra (>1600 compounds) and *inter alia* thermochemical data for over 7000 organic compounds.

Integrated Spectral Database
http://www.aist.go.jp/RIODB/SDBS/cgi-bin/cre_index.cgi

This spectral database for organic compounds is a free site organized by the National Institute of Advanced Industrial Science and Technology (AIST), Japan, and contains a growing collection of MS [EI-MS], FT-IR, ^{1}H and ^{13}C NMR, Laser Raman and EPR spectra.

Recommended sources (data collections, data bases and websites) for physico-chemical material properties: http://www.porperties.ethz.ch/en/

Bibliography

[1] D. Seebach, A. K. Beck, and A. Heckel. "TADDOL and its derivatives – our dream of universal auxiliaries," In G. Quinkert, M. V. Kisakürek, Eds., *Essays in Contemporary Chemistry: From Molecular Structure towards Biology*, Verlag Helv. Chim. Acta, *Zürich*, 2001; *Angew. Chem. Intl. Ed.* 2001, 40, 92–138.

[2] D. Seebach, M. Hayakawa, and J. Sakaki, and W. B. Schweizer, *Tetrahedron* 1993, 49, 1711.

[3] D. D. Ridley. *Information Retrieval: SciFinder® and SciFinder Scholar®*, John Wiley & Sons, Ltd, Chichester, 2002.

[4] S. R. Heller, Ed., *The Beilstein System: Strategies for Effective Searching*, American Chemical Society, Washington, 1998.

[5] R. E. Maizell. *How to Find Chemical Information*, 3rd edition, John Wiley & Sons, Inc., New York, 1998.

[6] L. M. Barnard. Structure Representation, In P. von R. Schleyer, Ed., *Encyclopedia of Computational Chemistry*, John Wiley & Sons, Ltd, Chichester, 5 Vols., 1998, 2818.

Links to patent documents:
EU: esp@cenet: http://ep.espacenet.com/
USA: USPTO: http://www.uspto.gov/patft/

Chapter 9

Laboratory Notebooks

An entry in a laboratory notebook constitutes the primary reference to any experiment one has personally carried out. It claims priority in case of any doubt. All relevant aspects of a conversion should be recorded, together with the order in which steps were carried out. All observations should be noted, in principle even those that at first sight appear unimportant. Only in this way can one ensure that the results of critical experiments can be reproduced.

Practical Organic Synthesis: A Student's Guide   R. Keese, M.P. Brändle and T.P. Toube
© 2006 John Wiley & Sons, Ltd.

At the beginning of each experiment record:

- the date;
- structural formulae (abbreviated, if necessary) and all reagents in order of addition;
- molecular formulae and molecular weights, preferably under the relevant structural formulae;
- literature references on the procedure (or on analogous preparations);
- weights (and number of moles) of each compound used;
- list of apparatus (with sketches in unusual cases);
- the purity of all compounds and solvents (which should have been determined!) *e.g.* 'Analar grade', 'freshly distilled from LiAlH$_4$', 'filtered through basic alumina, Woelm, activity I', 'single spot on tlc (SiO$_2$, hexane)', 'pure by NMR', *etc.*; and
- an assessment of hazards and safety precautions.

During the course of the experiment keep running notes on:

- all observations (described exactly);
- the order of individual operations and the time taken over each of them; and
- all experiments that have been used to keep a check on the course of the reaction.

At the end of the reaction record without delay:

- the method of working up;
- purification procedures;
- yields and percentage yields; and
- data on waste disposal.

Some suggestions

- Use a book with numbered pages.
- A book giving a carbon copy is a valuable safeguard!
- Write your report on one side of the page only, using the facing page for recording:
  weighings (unless a separate weighing book is used)
  distillation records
  titrations
  preliminary tests for separating mixtures
  hydrogenation curves
  suggested improvements for future use
  interpretation of spectra, *etc.*
- Note reference numbers of spectra in the margin (but keep the spectra themselves in separate files).
- Keep an index, which makes it easier to retrieve the information.
- For key experiments, write an edited report for later use and file it separately.
- Some research workers also keep a diary to provide a very brief record of what work they have done each day.

Sample pages from a laboratory notebook

LJ 4.23
20.04.05
p. 112

Reference: P. Renaud, C. Ollivier, P. Panchaud, Angew. Chem. Int. Ed. **2002**, 41, 3460.

Apparatus: 50 ml RB flask, septum, vacuum pump, nitrogen, magnetic stirrer, oil bath, Dimroth condenser.

Hazard Assessment:
- Ethyl iodoacetate: R25-34: poisonous if swallowed, caustic S26·36/37/39-45: wear lab coat, gloves, protective glasses; see a doctor if feeling unwell. Lachrymator: work in the fume hood.
- Phenylsulfonylazide: azides can decompose explosively at higher temperatures: use a protective screen.
- Hexabutyldistannane: R21-25-36/38-48/23/25: health hazard - avoid skin contact! R50-53: environmental poison. S35-60-61: dangerous waste! S45: see a doctor in case of accident or if feeling unwell.

Reagents: for a 10 mmol scale:
ethyl iodoacetate (2.140 g, 10 mmol)[Fluka, purum, ≥ 98%]
1-(2-methylenecyclohexyl)-nonan-3-one (4.728 g, 20 mmol)
phenylsulfonylazide (5.436 g, 30 mmol)
hexabutyldistannane (8.701 g, 15 mmol)[Fluka, pract., ≥ 95%]
di-tert-butyl hyponitrite (0.105 g, 0.6 mmol)

The olefin, azide, and iodoacetate were weighed out with the exclusion of air into a round-bottomed flask, previously heat dried and flushed with nitrogen. The flask was then flushed by 3 cycles of evacuation followed by N₂, and benzene (20 ml, dried by filtration under argon from alumina) was introduced. The

hexabutyldistannane was added via syringe and a Dimroth condenser attached to the flask. Half the initiator (3 mol·%)
was added and the apparatus placed in a 60°C oil bath.
After 2h the rest of the initiator was added and the reaction
continued for a further 2h. The mixture was then cooled
to room temperature and the solvent carefully removed using
a rotary evaporator. The crude yellow product was chroma-
tographed on silica gel (ca. 180g) containing potassium
fluoride (ca. 20g) using hexanes / ethyl acetate 9:1 as
eluent. The product fractions were concentrated at 40°C/200mbar.

LJ 4.23
20.04.05
p.2/2

Yield: 3.030g (8.45 mmol, 85% based on ethyl
         iodoacetate) as a yellowish liquid,
         ~40:60 mixture of diastereomers by NMR.

¹H-NMR 4.23

Separation: ca. 200mg of the diastereomeric
             mixture was separated by preparative
             HPLC on a Nucleosil 100-7 column
             using ᵗBuOMe /n-hexane 5:95 as eluent.

HPLC 4.23

Spectroscopic   IR : 2097 (N₃), 1736 (CO₂), 1716 (C=O) cm⁻¹.
Data :   ¹H-NMR (500 MHz, CDCl₃): major diastereomer:
                δ 4.09 (q, J = 7.1 Hz, 2H), 2.5-2.2 (m, 6H), 1.3-1.1
                (m, 24H), 0.83 (t, J = 7.0 Hz, 3H); minor diastereo-
                mer: δ 4.20 (q, J= 7.1 Hz, 2H), 2.55-2.3 (m, 6H),
                2.13 (t, J= 8.3 Hz, 2H), 2.0-1.8 (m, 2H), 1.8-1.7
                (m, 1H), 1.7-1.1 (m, 19H), 0.9 (t, J=6.1 Hz, 3H).
                ¹³C-NMR / DEPT: all expected signals present.

IR 4.23
¹H-NMR 4.23A
¹H-NMR 4.23B
¹³C-NMR 4.23A/B

Waste   • benzene and other organic solvents → yellow canister.
Disposal : • silica gel (contains tin and potassium fluoride)
            → container for solid toxic waste.

Bibliography

[1] H. M. Kanare. *Writing the Laboratory Notebook*, American Chemical Society, 1985.

Chapter 10

Writing a Report

A good clear report is easy to produce if one has a comprehensive description of the work including all relevant data on the starting materials and products as well as all the experimental details in one's laboratory notebook. The experimental procedure should be described concisely with neat formulae and relevant references. If a reaction has been carried out several times, at least one of the experiments should be described fully. All the analytical data quoted should be for the experiment described, unless otherwise indicated.

A report on a preparative experiment should have the following features:

Title
This should include the name of the product, the names of the experimenters, and, where relevant, the date, *e.g.* 'Isolation of cyclopenten-3-one from....', 'Synthesis of cyclopenten-3-one from...'

Report
This can be arranged in the following sections:
(a) *Method*: Here the overall transformation carried out in each step of a multistage synthesis is described, *e.g.* 'Pinacol is prepared by the reductive dimerisation of acetone', '*endo*-Bicyclo[2,2,1]hept-2-ene-5-carboxylic acid is formed by the cyclo-addition (Diels-Alder reaction) of cyclopentadiene and acrylic acid'.

(b) *Reaction Scheme*: This shows the transformation of starting materials to products by means of formulae (configurational or conformational, if necessary). Reaction conditions (reagents, solvents, catalysts, temperature, *etc.*) are indicated in

Practical Organic Synthesis: A Student's Guide   R. Keese, M.P. Brändle and T.P. Toube
© 2006 John Wiley & Sons, Ltd.

abbreviated form above and below the arrows in the usual way. The molecular formulae and molecular weights can be appended to the relevant structures. All structures should be numbered. In general, diagrams showing mechanisms are presented separately.

(c) *Experimental Section*: The description of the experiments (past tense, third person and passive voice) should be sufficiently detailed to permit the repetition of the reaction without further consultation of the literature. The report should be sufficiently complete for it to be used in preparing a paper for publication. The weights of all compounds (and the number of moles), the purity of starting materials and solvents* and all relevant reaction conditions (temperature, time, pressure, *etc.*) should be quoted, as well as the work-up and purification procedures. One should also provide information about the apparatus used, any peculiarities observed and simple procedures for following the course of the reaction.

In the text, names of all chemicals should be written out in full; formulae are used only in reaction schemes. By contrast, abbreviated or trivial names (with the structure numbers used in the reaction schemes) make it easier to follow descriptions when long and complex names are involved.

The yield is quoted (*not* the average yield over several preparations) with an indication of purity ('crude', 'after recrystallisation', *etc.*), as well as the literature yield with references (where relevant).

Finally, the physical data used to characterise the compound should be reported (again with literature references): mp, bp, n_D (with temperature superscript), R_f (with details of tlc system), IR, UV, NMR, MS, *etc.*

(d) Additions and results from other experiments and suggestions for further experiments: This section is necessary if there is no general discussion of the reaction, if variation in any reaction parameter leads to widely different results, if the method of work-up is critical, or if alternative attempts to isolate the product have failed.

Some typical expressions and abbreviations
bright yellow crystals (10.5 mg, 78% based on (8))
tetramethylsuccinic anhydride (33.8 g, 0.217 mol)
nitrile (1.15 g, 8.5 mmol)
a solid residue (68.9 g) remained, which was recrystallised from methyl acetate (*ca.* 250 cm³) with charcoal decolorisation
in absolute ethanol (2 cm³)
poured onto ice (1.5 kg)
in sodium hydroxide solution (1 M)
with methyllithium in ether (1.49 M, 16 cm³)

---

* These data on substances used repeatedly in a series of experiments can be collected and placed at the beginning of the experimental section, if desired.

after addition of hydrochloric acid (18% solution, 400 cm^3)

3 s

4 min

2.5 h

50–60 °C

the fraction bp 179–195 °C/2 Torr

a sample subliming at 110 °C/0.005 Torr

in a 100 cm^3 round-bottomed flask fitted with reflux condenser

tetramethyllaevulinic acid (**13**) (9.15 g, 53 mmol) was converted into the acid chloride (**14**) by treatment with thionyl chloride (6 cm^3)

oxidation with Fetizon's reagent[4].

Bibliography

[1] For a listing of the journal abbreviations and the complete journal titles of approx. 1300 key chemical journals see http://www.cas.org/sent.html

[2] A short list of *ca.* 50 journals and their abbreviations is available: http://www.elicaps.ethz.dv/en/practical.html

[3] The main source for the current abreviations and bibliographic details of all journals covered by CAS is the CAS Source Index (CASSI), available in print format, as CD-ROM or on a local server with a licence.

[4] H. F. Ebel, C. Bliefert, and W. E. Russey. *The Art of Scientific Writing: From Student Reports to Professional Publications in Chemistry and Related Fields*, 2nd edition, John Wiley & Sons, 2004.

[5] B. Gustavii. *How to Write and Illustrate a Scientific Paper*, Cambridge University Press, 2003.

[6] J. S. Dodd, Ed., *The ACS Style Guide: A Manual for Authors and Editors*, 2nd edition, American Chemical Society, 1997.

[7] V. Booth. *Communicating in Science: Writing a Scientific Paper and Speaking at Meetings*, 2nd edition, Cambridge University Press, 1993.

Chapter 11

Example of a Laboratory Report

1-Phthalimido-trans-2,4-diphenylaziridine

12 December 2005 John Smith

Practical Organic Synthesis: A Student's Guide   R. Keese, M.P. Brändle and T.P. Toube
© 2006 John Wiley & Sons, Ltd.

(a) Method
trans-Stilbene   reacts   with   N-aminophthalimide   in   the   presence   of
lead   tetraacetate   in   an   oxidatively   induced   (2+1)   cylcoaddition   to   yield
1-phthalimido-trans-2,3-diphenylaziridine(1,2).

(b) Scheme

$C_{14}H_{12}$ $C_8H_6N_2O_2$ $C_{22}H_{16}N_2O_2$
mw: 180   162   340

(c) Experimental
N-Aminophthalimide* (6.50 g, 40 mmol) and trans-stilbene** (36.0 g, 200 mmol)
were vigorously stirred in dry dichloromethane† (100 mL) in a three-necked 500 mL
flask fitted with a Teflon-bladed stirrer, a thermometer, and a powder funnel. Lead
tetraacetate†† (20.0 g, 40 mmol) was added at r.t. to the suspension over 10 min. After
further 30 min stirring, the mixture was filtered through Celite and concentrated on
a rotary evaporator at 40°C. The crude product was at once transferred to a silicagel
gel (190 g) column and the excess stilbene eluted with dichloromethane. A second
fraction containing a small amount of an unidentified byproduct was then eluted, followed by 1-phthalimido-trans-2,3-diphenylaziridine (10 g).

The product was recrystallised overnight at 0°C from chloroform/pentane and the
yellow needles dired for 2 h at 25°C/0.05 torr. The yield of the recrystallised material
m.p. 177–179°C was 5.48 g . A second crop of crystals (4.44 g), m.p. 176–177°C was
obtained from the mother liquor.

Yield after chromatography: 10.0 g (73.5%)
Yield after crystallisaton: 9.92 g (73.0%)

(d) Physical data
R_f (silicagel, dichloromethane): 0.44
m.p. 177–179°C (lit [1] m.p. 165°C)
IR (chloroform), vmax: 1774 (m), 1718 (s) cm^{-1}
^{1}H-NMR (CDCl3, 60 MHz): 3.96 and 4.97 (AB, J=6 Hz, 2 H), 7.08-783 (m, 14 H)
ppm

---

* Fluka, purum, mp 199–202 °C.
** Fluka, puriss.
† Distilled over phosphorus pentoxide.
†† Fluka, purum, 85–90%, moistened with acetic acid.

MS (m/z): 340 (M$^+$, 9%), 194 (100)

Anal. calc. for $C_{22}H_{16}O_2N_2$ (340.) C 77.63, H 4.74, N 8.23; found C 77.55, H 4.79, N 8.29

(e) Notes

If the filtrate obtained by filtration of the reaciton mixture trhough Celite is treated with an equal volume of pentane at 0°C, instead of being concentrated, a crystalline precipitate is produced. Recrystallisation from pentane/dichloromethane gives a yield of 55–65%.

Heating of the reaction mixture at 100°C for 15 h did not affect the result (IR, tlc).

(f) References

[1] L. A. Carpino and R. K. Kirkley, J. Am. Chem. Soc. 1970, 92, 1784.
[2] D. J. Anderson, T. L. Gilchrist, D. C. Horwell and C. W. Rees, J. Chem. Soc.(C), 1970, 576.
[3] Fluka, puriss.
[4] Fluka, purum, m.p. 199–202°C.
[5] Distilled over phosphorous pentoxide
[6] Fluka, purum, 85–90%, moistened with acetic acid.

Some Quotations from Laboratory Reports

The gradual depression of the melting point leads one to the conclusion that, far from the 3-indolylacetic acid increasing in purity, the sample was in fact being enriched in the impurities.

condition.[6] As after the third experiment no reasonable cause remained to account for the decomposition as all possible sources of adventitious moisture had now been eliminated, the dream of a pheromone synthesis was buried with a heavy heart and an alternative experiment undertaken.

As the solid could not be filtered off by any of these techniques, it was sedimented in the ultracentrifuge. The aqueous supernatant contained brown oil and white solid bodies. The desired product was believed to be these white bodies, so they were washed twice with water. This treatment, however, revealed that they were only common salt, and that the substance was in fact the brown oil. This error naturally led to severe losses.

It seemed probable that the compound had been produced. The extraction with bicarbonate solution was therefore repeated, using ethyl acetate in place of ether, but again no product was obtained.

evaporated in order to allow the benzalacetophenone to crystallise. This did not, however, occur. An analogous attempt using alcohol failed equally. The only remaining alternative seemed to be

saturated brine and extracted with ether (3 x 100 cm³). After distillation of the ether the residue (ca. 0.5 cm³) was subjected to short path distillation. This procedure produced no result, however (no yield).

hydrolised. The large loss on the first recrystallisation was caused by the use of a long-necked flask.

Of the 2nd substance very little was eluted as the column had run dry overnight for some reason.

yield: 7.025 g (0.047 mol, 31%) (lit. yield 87%).

The poor yield may be ascribed to the use of an unsuitable distillation apparatus.

Chapter 12

Hints on the Synthesis of Organic Compounds

12.1 THE ROUTE TO SUCCESS

Preparative organic chemistry is a variable amalgam of science, art and craft, with the quest for new compounds or conversions as its goal. It is often difficult (and sometimes

Practical Organic Synthesis: A Student's Guide   R. Keese, M.P. Brändle and T.P. Toube
© 2006 John Wiley & Sons, Ltd.

impossible) to bring about a desired transformation. As a result, between one's original concept and the joy of success lie deserted acres of vain endeavour and disappointment. Things can be very tough!

The 'Hints on Organic Synthesis' given below should not be considered as a recipe for certain success if followed rigidly, nor yet as a comprehensive list of errors. They are rather a contribution, mostly the fruit of bitter experience, to the more effective organisation of research work.

The aim of an organic synthesis is the preparation in one or more stages of a particular organic compound. Alternative routes will always present themselves: A great deal depends on choosing the best one. The section on database searching (Chapter 8) as well as the works listed in the Bibliography aid this choice.

What criteria govern this choice?

• Availability of starting materials and reagents (commercial products? delivery times!? given in *Organic Syntheses?*).
• Contribution to chemical knowledge.
• Dangers (poisonous? explosive? inflammable?).
• Amount of time and labour required.
• Cost.
• Availability of good literature preparations.
• Selectivity.
• Environmental effects. The increasing awareness of the need to protect the environment means that one should develop and use methods with high selectivity, producing no side-products and the smallest possible amount of waste.

Once one has decided on a synthetic route, one needs to sort out the following matters:

Procedure
If a literature procedure is available, read it carefully (including footnotes, which in, *e.g. Organic Syntheses* may contain vital information). If the preparation is not reported in the literature, it may be possible to modify published procedures (*e.g.* in *Organicum*) or syntheses of related compounds. As far as possible one should endeavour to understand the reaction pathway. Careful comparison of the literature reports on related conversions will often indicate which reaction parameters are critical, where crucial phases in the reaction lie and which side-reactions are to be avoided. Creative geniuses can use their knowledge, intuition and experience to devise experimental procedures which are completely novel – with appropriate precautions! For information on procedures see also the databases CrossFire and Scifinder Scholar® as well as the review literature cited in Chapter 8.3.

Plan of Work
The stages to be considered are reaction, work-up, purification and characterisation.

Develop alternative solutions and weigh them up against each other.

Assess the hazards associated with the planned reactions.

In order to choose the optimal conditions for work-up and purification, the properties of the product should be known or at least estimated (state of aggregation, solubility – low-MW alcohols, amines, aldehydes and ketones are often water soluble) mp, bp, reactivity, (e.g. towards water). Consider whether the reaction is exothermic or endothermic. With large quantities, exothermic reactions may become too hot! In such cases initial heating will need to be carefully controlled, and arrangements should be made for rapid removal of heat and application of cooling if it should become necessary.

Planning the Use of Time
An estimate of the actual duration of each step in the procedure is an advantage. In particular, attention should be directed to establishing at which stages the process could be interrupted if necessary. Beware of a tendency to underestimate the time taken to work-up.

Scale
It is important to estimate this. Choose a scale that makes handling easy. A useful rule of thumb: Use enough starting material to give a theoretical yield of 10–20 g. In many cases the preparation will need to be repeated several times. Experience shows that the second preparation usually goes better than the first!

Preliminary Work
Check availability of all materials (stores, reagent shelves, orders, *etc.*). Purify reagents and solvents, and set up apparatus. Check all chemicals required for identity and purity (appearance, smell – but do not inhale, mp, bp, n_D, NMR, GLC, tlc *etc.*). Make sure you know the methods for destruction of excess active reagents (alkali metals, catalysts – pyrophoric when filtered, hydrides and toxic materials) (see Chapter 13).

Preliminary Tests
It is a good idea to try out all reactions on about 1/10th of the planned scale. For completely new reactions, even smaller scale tests are desirable in order to see if the reaction takes place at all.

Apparatus
Heterogeneous mixtures greater than 10–20 cm^3 require adequate mechanical stirring to reduce concentration and temperature variations. Magnetic stirrers are appropriate for homogeneous solutions.

Reaction vessels should be of adequate size (*i.e.* never more than $1/2$–$2/3$ full). The only sealed apparatus to be used are autoclaves and ampoules: Unwittingly sealed apparatus is a classic cause of accidents! If air or moisture have to be excluded, use a drying tube or an inert gas stream vented through the valve shown on p. 128.

Laboratory Notebook
Note the hazards and control measures needed for starting materials and products (see Chapter 2). Follow the procedures for keeping a laboratory notebook (Chapter 9).

Reaction
In order to have a reproducible reaction, one must have *well-defined reactions parameters* (concentrations; temperature – thermometer in the reaction mixture; exclusion of oxygen, moisture, light, *etc.*; reaction time). If possible, one should monitor the course of the reaction by some simple technique (tlc, pH, appearance of a precipitate, spot tests, *e.g.*, KI/starch paper for starting materials or product, spectroscopy, GLC). Of course, the 'well-defined reaction parameters' should include an *unambiguous termination to the reaction* (without uncontrolled further reactions)! This may be achieved by cooling, dilution, removal or decomposition of reagents, etc.

Work-up and Isolation
The method of working up a reaction mixture depends on the relative properties of the product and of any other materials present. One should always consider all the alternatives and choose the most efficient one.

In nine cases out of ten the following standard method will do:
Pour onto ice, extract with solvent, wash organic layer (neutral/acidic/basic as appropriate), dry, evaporate solvent (rotary evaporator for ether, dichloromethane, pentane – use a manustat with a safety valve and water pump, or a membrane pump with an efficient condenser to reduce water consumption and contamination of the water supply, especially when using dichloromethane. For products of low bp distillation of the solvent via a Vigreux column is often advantageous). Weigh the crude product.

Now ask the *key question*: Is this the desired compound? *Yes!*; *no!*; or more often: *perhaps!!*

Use IR/NMR/TLC/GLC to test its purity, and with luck to give an unambiguous answer to this question.

Possible Scheme for an Organic Transformation

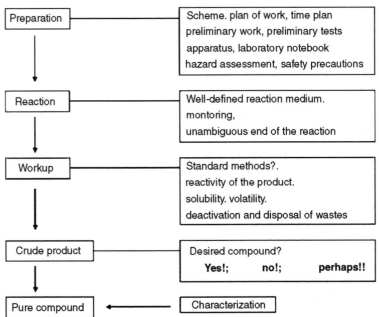

Purification
Because many substances are very unstable when impure, a crude product should not be left any longer than necessary. Decomposition can often be delayed by storage at low temperature and/or in the absence of moisture and light, or by the addition of stabilisers.

If the method of purification has not already been established, use only portions of the crude product to try out various techniques. Decomposition often occurs during purification operations.

In practice, the transfer of a purification technique to one of higher capacity (tlc → preparative tlc or column chromatography, bulb distillation → column distillation) often produces complications.

Principle
Always try the simplest procedures first, even if they appear to have little hope of success. Once the substance is completely pure it should be characterised at once. Even well-known compounds require some characterisation (mp, bp, IR, tlc, n_D, GLC etc.). One should obtain one value for identity and another for purity.

A small sample of each pure stable substance should be retained in case further information is needed. For important substances that are difficult to crystallise, the total amount of crystalline material should never be redissolved for recrystallisation: Always save some seed crystals!

Disposal and Recovery of Solvents
Reagents and catalysts must be deactivated before disposing them of (see Chapter 13). Halogenated substances and solvents must be kept separate from nonhalogenated ones. Large quantities of solvents may be recovered by distillation.

Optimisation of Reaction Conditions
Make a list of all the factors that could influence the course of the reaction (with some idea of how much they might vary).

Then try the reaction under the conditions you judge to be most favourable. If this attempt does not produce any improvement, try varying the above factors one at a time to a considerable extent (for optimisation it is essential to vary only one parameter at a time, otherwise it will not be possible to say what caused an improvement or deterioration). Try to use an economical method for estimating the yield.

- Isolate and characterise by-products. One can often glean useful information in this way.
- Results are the consequence of perseverance coupled with ingenuity. Sheer obstinacy seldom leads to success.

12.2 SOME TIPS ON HANDLING WATER- AND AIR-SENSITIVE SUBSTANCES

The manipulation of air- or moisture-sensitive substances is now a common part of general laboratory practice. The following basic procedures will cover many situations, but will need to be modified considerably for special cases. Between passing a stream of nitrogen through a reaction mixture and working in a dry-box filled with an inert gas containing only a few ppm of oxygen or on a vacuum line, there are many stages. The simplest aids for handling air- or moisture-sensitive compounds are characterised by being able to be applied when needed, merely amplifying existing glassware. The following 'tips' are to be seen as suggestions that can lead to improved technical solutions.

12.2.1 Simple Operations Under Inert Gas [1]

Which protective gas to use

The most common protective gas is nitrogen, with argon as next choice.

Nitrogen

This is available in various purities. For many practical applications one needs a quality of 99.995%, although this still contains 10 ppm of oxygen and less than 10 ppm water.

Argon

In those cases where N_2 can react (e.g. in the preparation of lithium sand, or with complexing transition metals), argon should be used. Again, different qualities are available and for most applications one should use 99.998%, which contains less than 2 ppm of oxygen and less than 3 ppm of water. Argon has an advantage over nitrogen in that, on account of its greater density, argon-filled vessels can be opened briefly without the need for additional precautions provided there is no turbulence.

PVC and rubber tubing are permeable to oxygen, so it is advisable to keep passing through a stream of protective gas for the duration of the reaction.

How to prepare the reaction vessel?

Ideally, the reaction vessel should be heated overnight at 125 °C in a drying oven together with any requisite dropping funnels, stirrers, condensers, *etc.*; the whole apparatus is assembled while still hot and then allowed to cool under protective gas. In some cases it is sufficient to heat the apparatus carefully with a hairdrier while passing protective gas through it.

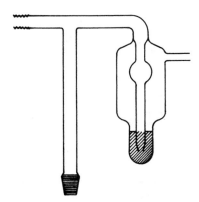

When passing gas over or through a solution the valve shown in the figure has been found ideal. It allows one to admit the gas and at the same time check its flow. Reaction vessels fitted with such a valve can be opened briefly because a counter-flow of inert gas is formed. A pressure-equalising dropping funnel is particularly useful as the inert gas will then also protect the contents of the funnel.

How to degas solvents?

The simplest procedure is to boil the solvent for some time while passing through the inert gas and then to distil in an inert atmosphere.

Small quantities (5–10 ml) of a solvent, as long as it is not too volatile, can also be freed from interfering gases as follows: Place the solvent in a flask fitted with a ground glass tap, cool to −70 °C, and evacuate the flask using a high-vacuum pump. Warm to room temperature. Repeat the process twice.

One can also degas solvents by placing them in an ultrasonic bath for 10–15 minutes, under mild vacuum if necessary.

How to transfer small quantities (< 20 ml) of oxygen- or water-sensitive liquids?
For this purpose it is particularly convenient to use a syringe with a needle (canula) whose length may reach that of a normal NMR tube. The syringe and needle are dried at 125 °C in a drying cabinet and allowed to cool under protective gas, perhaps in a desiccator. Stock bottles and reaction vessels are fitted with serum caps and solutions are transferred by syringe. Ideally, the material removed should be replaced by an inert gas admitted using a syringe needle and bubble valve. If it is not possible to equalise pressure with inert gas, one can compensate by filling the syringe – before piercing the serum cap – with a quantity of inert gas equal in volume to the liquid to be withdrawn. For larger quantities one could use a dry pipette fitted with a detachable 'propipette'.

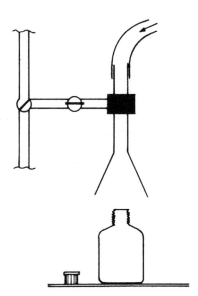

For the transfer of larger quantities of oxygen- or water-sensitive solutions or liquids, long, flexible stainless steel syringe needles (canula) are used; the needle is introduced via a septum into the gas space in the vessel above the sensitive liquid and is flushed using a slight excess pressure of protective gas. Then the free end of the needle is inserted via another septum into the receiving flask, which is equipped for the passage of protective gas. The end of the needle in the vessel containing the sensitive liquid is then pushed below the surface of the fluid: The slight excess pressure causes the liquid to be transferred.

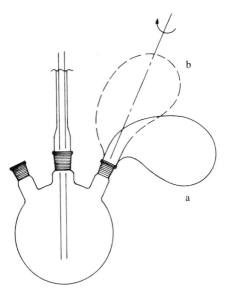

Figure 12.1: Apparatus for transferring a moisture-sensitive solid

Reagent bottles with screw tops may conveniently be opened under an inverted funnel through the stem of which protective gas is being passed (see Figure on page 132), or in a plastic bag flushed with gas. With higher boiling solvents, such as THF, a U-shaped glass tube with protective gas passing through it can be hung over the neck of the bottle.

How to transfer solid substances which are moisture sensitive?
The easiest way to transfer solids is to follow the procedure described above for re-agent bottles with screw tops. To add a moisture-sensitive solid in portions to a reaction mixture a flask, constructed as in Figure 12.1 with a ground glass joint, is ideal. It is important that the shape is chosen to ensure that in position (a) no material can trickle into the reaction mixture, while in position (b) all the remaining material can be tapped lightly into the reaction flask, making sure that even in position (b) the flask does not get in the way of a condenser or stirrer fitted to the reaction vessel.

In place of such a flask one can successfully use an ordinary flask connected to the reaction flask by a flexible tube of large internal diameter.

How to filter under protective gas
The 'pressure filter' shown in Figure 12.2 has proved extremely suitable for this purpose. In order to filter, one introduces inert gas through (a) and then inserts a loose-fitting plug at (b). If one is mainly interested in the filtrate, it is advisable to use a two-necked flask fitted with a ground glass tap – flushed with inert gas before the sinter is removed – or to fit between the sinter and the flask a straight adaptor through which protective gas can be passed when the sinter is removed. Sensitive suspensions can be transferred directly into the pressure filter from a flask by using a tap fitted with ground glass cones at each end. For filtration of a reaction mixture under protective gas, see also [2].

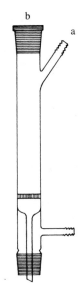

○ Figure 12.2: Pressure filter

In less critical cases filtration can be carried out under protective gas in a conventional 'glove bag' [3].

How to crystallise air- or moisture-sensitive substances
For this purpose the filter with the side tube at the top, as shown in Figure 12.3, is suitable [4]. A collecting flask can be attached to the apparatus. The sensitive material

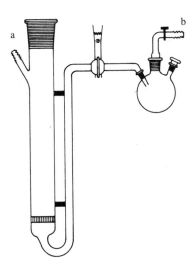

○ Figure 12.3: Apparatus for crystallising air- or moisture-sensitive substances is introduced as a suspension and, as in the case of the pressure filter described above, is separated from the solvent by forcing protective gas at moderate pressure through (a). To crystallise the material, gas is then passed in through the collecting flask at (b), serving, in addition, to stir the solution when the solvent is added. If required the whole filter can be immersed in a cold bath.

In the simplest cases, crystallisation under protective gas can be carried out in a two-necked flask fitted with a ground glass tap which serves to introduce the gas. In this case the mother liquor will need to be removed by pipetting or siphoning.

How to store moisture- or air-sensitive substances
For this purpose one may use a two-necked flask fitted with a serum cap and a ground glass tap. Ampoules fitted with serum caps from which fluids and solutions can be removed when needed using a syringe are better for longer term storage. However, for the best long-term results, ampoules sealed under inert gas or after evacuation are often essential.

To seal ampoules successfully, the ampoule neck needs to have been constricted beforehand. Cool the contents adequately, evacuate, close the vacuum stop-cocks and seal using a fine flame.

Manipulation of some commonly used reagents

Lithium Alkyls, Diborane in THF, Diethylaluminium Cyanide in Toluene, Diisobutyl-aluminium Hydride in Hexane
For transfer from the usual commercial bottles (with screw tops) see Fig. on page 133.

Lithium Aluminium Hydride
The powdered reagent is extremely sensitive to moisture. It is advisable to open the can in which it is supplied under an inverted funnel through which protective gas is being passed and to weigh it out only in closed vessels. Before closing the stock bottle flush it with protective gas. Once opened, the cans should have their plastic lids firmly replaced and should then be stored in a desiccator filled with protective gas. It may be more convenient to use the commercially available 1 g tablets: In crystalline form the reagent is not pyrophoric even in moist air. LiAlH$_4$ may also be obtained as solutions or suspensions in various solvents [5].

Potassium Hydride
The powdered reagent is usually supplied commercially under paraffin oil. The sedimented powder is pipetted out, under protective gas, using a pipette with a wide orifice. The adhering paraffin oil can be removed by filtration and washing with dry, oxygen-free pentane, using a pressure filter (See fig. 12.2). Dry, oil-free potassium hydride flows easily, but in the presence of moisture is spontaneously flammable. It is advantageous to use a preweighed sinter so that one can determine the weight of dry reagent.

Lithium Sand
Powdered lithium in a dry and flowing condition can be prepared as follows:
In a cylindrical vessel fitted with a vibratory stirrer, an argon inlet, and a drying tube
(see Figure 12.4), lithium wire (2–5 g) and the 2% of sodium needed for the preparation of lithium alkyls are heated in boiling dry tetralin (bp 207 °C, 50 ml). The molten
lithium is stirred vigorously and then the heating and vibrator are switched off and the
suspension is cooled. The tetralin is carefully covered with a layer of dry pentane; the
light lithium rises to the top of the pentane layer. Using a pipette most of the brown-
ish tetralin solution beneath the pentane is removed and the lithium sand is washed
several times with more pentane. To obtain dry lithium, the pentane is removed by
filtration under pressure of protective gas in a pressure filter (See Fig. 12.3). Protective
gas is passed through the lithium sand to dry it. Lithium may be obtained as a dispersion in mineral oil or as powder [6].

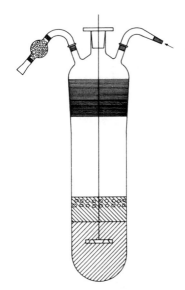

Figure 12.4: Apparatus for preparing lithium sand

Sodium sand
See, *e.g. Org. Synth., Coll. Vol. V*, p. 1090.

Potassium
See, e.g. *Org. Synth., Coll. Vol. IV*, p. 134. Potassium cannot be freed from adhering
oil by the usual drying paper, as it may spontaneously catch fire in the process. Potassium cut under liquid paraffin or xylene should be dipped briefly into dry hexane and
immediately transferred to a reaction vessel flushed with protective gas.

12.3 HINTS ON WORKING AT LOW TEMPERATURES

12.3.1 Increasing Reaction Selectivity

At lower temperatures, the formation of by-products is often reduced and unstable compounds and reagents can be handled with greater safety. Reactions which lead to two or more similar products (*e.g.*, isomers) almost always give altered product ratios at different temperatures, although care is needed to ensure that the temperature is not so low that it would take an unacceptable time to achieve a reasonable degree of reaction. One can monitor this process either by measuring the temperature rise when reagents are added or by tlc or GC analysis of the progress of product formation. Addition of a large quantity of a highly reactive reagent below the temperature at which reaction will take place can delay the onset of an exothermic process in a manner that permits considerable control.

It is more expensive in terms of energy and apparatus to carry out reactions at low temperature than at room temperature. Technical and economic considerations mean that it is reasonably feasible to work at $-20\,°C$; lower temperatures are more difficult. The fact that working at lower temperatures has advantages in terms of yield, product purity and simpler purification processes, reduced amounts of unwanted by-products requiring disposal, and increased safety has to be set against the increased expense [7].

12.3.2 Recrystallisation

Many compounds which melt below room temperature can be crystallised without difficulty using the apparatus, shown on p 133 (Figure 12.3), leading to appreciable purification in a manner which often complements that achieved by chromatography or distillation. In general, it is possible to remove traces of solvent from these samples after repeated recrystallisation by a simple vacuum distillation (but it is advisable to test for thermal stability on a small sample first, checking the distillate by GC or HPLC!).

Hints

Preparation of seed crystals	(see Chapter 3)
Particularly suitable solvents	methanol, ethanol
(check mp)	diethyl ether
	pentane (*iso*-pentane below $-160\,°C$)
	and mixtures (also for lowering viscosity)

Purity
Analyse the melted crystals and the mother liquor by GC or HPLC in order to ascertain the efficiency of the crystallisation procedure.

Temperature control

Desired temperature	Method of cooling [8], [9]
above *ca.* +80 °C	air (relatively inefficient, as it requires large exchange volumes)
ca. +15 to 80 °C (e.g. for distillates which solidify above the temperature of the cooling water)	water (possibly in a large reservoir), circulating pump, thermostat
ca. +5 °C	ice/water (0 °C)
+5 to *ca.* −10 °C	NaCl/ice mixtures (lowest temperature −21.1 °C with ice: salt = 10:3)
to *ca.* −20 °C	$CaCl_2·6H_2O$/ice mixtures (lowest temperature *ca.* −30 °C with ice: salt = 1:1)
to *ca.* −70 °C	Solid CO_2 (sublimes at −78.5 °C) in a slush bath with *iso*-propanol, acetone or methanol.
−195.8 °C	Liquid N_2 Evacuate apparatus before using a liquid nitrogen cold trap as otherwise liquid O_2 (bp −183.0 °C), which reacts explosively or pyrophorically with organic compounds, may be condensed.

Electrical thermostat are obtainable with working ranges from +250 to −50 °C.

Cryostatic slush baths
Constant, reproducible low melting temperatures can be attained with pure compounds using Dewar flasks. One needs to ensure that the solid and liquid components are thoroughly mixed, e.g. by generating the solid component by carefully adding liquid nitrogen. (Take care to avoid crust forming and, especially with glass Dewar flasks, the possibility of dangerous implosion. *Safety spectacles are essential!*)

Solvents for slush baths	mp [°C]	Solvents for slush baths	mp [°C]
cyclohexane	6.5	ethyl propionate	−73.9
water	0.0	ethyl acetate	−83.6
tert-amyl alcohol	−12.0	acetone	−94.6
benzyl alcohol	−15.3	*iso*-amyl alcohol	−117.2
pentan-3-one	−42.0	methylcyclohexane	−126.3
diethyl malonate	−50.0	*iso*-pentane	−160.5

12.4 HINTS ON THE SYNTHESIS AND ANALYSIS OF CHIRAL COMPOUNDS

Methods for preparing enatiomerically pure chiral compounds are important not only in the fields of pharmaceuticals, fragrances and the synthesis of other physiologically active substances but also in fundamental research [10]. Historically, the preparation of racemic products was frequently acceptable, but it is now increasingly important that chiral materials are synthesised with high enantiomeric purity. The following section outlines some of the ways in which optically active molecules can be obtained and the most important methods for measuring optical purity [11]–[13].

12.4.1 Methods for Preparing Chiral Compounds

(a) *By resolution of racemates* [14]
The racemate is treated with a single enantiomer of an auxiliary chiral material to produce a mixture of two diastereomers, *e.g.* a racemic carboxylic acid can be converted into the respective diastereomeric amides or amine salts by reaction with a chiral amine. The diastereomers can then be separated by fractional crystallisation or chromatography, which, following cleavage of the chiral auxiliary, yields the required two enantiomers (ideally in optically pure form).

The resolution of racemates is often rather time-consuming and, if one does not require both enantiomers for further use, it implies a loss of at least 50% of the material. In some cases, they can be separated by preparative column chromatography using a chiral phase [15,16]. However, it is generally preferable to prepare the desired enantiomer using one of the methods (b) or (c), given below.

(b) *Using a non-racemic chiral starting material* [17]
There are now many chiral starting materials offered for sale at high enantiomeric purity; the specialised catalogues of suppliers provide information about their uses. There are also many review articles dealing with the synthetic applications of such materials, especially those from the naturally occurring 'chiral pool' [17c].

(c) *By asymmetric synthesis from non-chiral starting materials and by asymmetric catalysis*
Many efficient methods for synthesising chiral compounds with high enantioselectivity from achiral starting materials have been reported in recent years. These frequently involve the linking of an achiral synthon to a chiral auxiliary, leading to a diastereoselective reaction under the influence of the chiral auxiliary. The chiral auxiliary is removed in a later step and, ideally, recovered. Chiral reagents and catalysts which permit the conversion with high enantioselectivity of achiral starting materials into chiral products are becoming increasingly important. Such chiral methods have been extensively reviewed [18,19].

12.4.2 Determination of Enantiomeric Purity

The common measure of the enantiomeric purity of a compound is the enantiomeric excess (ee%), which is defined as follows:

$$ee(\%) = \left\{ \frac{\left[(x-y)\right]}{\left[(x+y)\right]} \right\} \cdot 100$$

where x, y = relative amounts of the two enantiomers.

The relative amounts of the two enantiomers can be determined by chromatography or NMR spectroscopy. The optical purity measured by polarimetry is not always identical with the ee and should thus be used as a measure of enantiomeric purity only with great circumspection.

(a) *Chromatographic methods* [17a]
The compound for analysis is linked to an enatiomerically pure chiral molecule, generating a pair of diastereomers whose ratio can be measured by GC or HPLC. It is essential that derivatisation proceeds quantitatively, otherwise one of the diastereomers might be enriched by kinetic discrimination; one can check whether this has occurred by repeating the process using the other enantiomer of the chiral reagent. Because the determination relies on the enantiomeric purity of the chiral reagent, its ee needs to be known accurately. One can also in many cases separate the untreated mixture of enantiomers directly on a chiral GC or HPLC column.

Of all the methods of determining enantiomeric purity, analysis by GC or HPLC is the most accurate, especially for samples of very high ee, as the presence of amounts of the minor enantiomer even less that 1% can be detected reliably.

(b) *NMR spectroscopy*
NMR signals of a pair of enantiomers can be shifted differentially using a chiral shift reagent (most commonly a europium complex). If it is possible to separate the two signals for corresponding nuclei in the two enantiomers to regions where they do not overlap with any other signals, the ratio of the enantiomers can be determined by integration. If it is not possible to separate the signals in this way, one can prepare diastereomeric derivatives (as for the chromatographic methods above) and measure their ratio by integration, e.g. enantiomeric alcohols can be esterified using either (R)- or (S)-2-methoxy-2-trifluoromethylphenylacetic acid in the presence of dicyclohexylcarbodiimide and the diastereomeric mixture analysed using ^{1}H, ^{13}C or ^{19}F NMR spectroscopy.

(c) *Polarimetry* [17a]
The specific rotation $[\alpha]_\lambda^T$ of an optically active compound is given by:

$$[\alpha]_\lambda^T = \frac{\alpha}{l \cdot c}$$

where $[\alpha]_\lambda^T$ = specific rotation (°) of the compound at wavelength λ (nm) and temperature T (°C). In general λ (nm) corresponds to the sodium D line = 589 nm

α = measured rotation

l = cell path length (dm)

c = concentration (g/ml)

The solvent and concentration should also be specified, *e.g.*, $[\alpha]_{589}^{23}$ or $[\alpha]_D^{23} = -19.8$ ($c = 0.843$ g/ml; $CHCl_3$)

The optical purity p of a compound is given by:

$$p\,(\%) = \frac{[\alpha]_\lambda^T}{[\alpha_E]_\lambda^T} \cdot 100$$

or else: $p\,(\%) = \left([\alpha]_\lambda^T / [\alpha_E]_\lambda^T\right) \cdot 100$

where $[\alpha_E]_\lambda^T$ = specific rotation of the pure enantiomer

Because the concentration dependence of optical rotation is not strictly linear, this method is only reliable when rotations are measured under strictly comparable conditions of solvent, concentration and temperature.

In general, measurements of optical purity using ee and p agree, *i.e.* ee (%) = $p\,(\%)$. However, in some cases, especially where the compounds can associate (*e.g.* carboxylic acids), the two values may differ perceptibly. Measurements of enantiomeric purity by measuring optical rotation can also be compromised by, *e.g.* chiral impurities in the sample or solvent. In addition, the specific rotation of the pure enantiomer has often never been determined.

The ee value for a particular sample should, therefore, always be checked using an alternative method.

12.5 HINTS ON THE SYNTHESIS OF ISOTOPICALLY LABELLED COMPOUNDS

12.5.1 Synthesis with Stable Isotopes

Compounds containing isotopes in other than the proportions in which they occur naturally may be required, *e.g.* as NMR solvents or for the elucidation of reaction mechanisms.

Isotopically labelled target molecules may in most cases be prepared from simple starting materials that may be obtainable with different degrees of isotopic enrichment:

^2H	^{12}C	^{13}C	^{15}N	^{17}O	^{18}O
D_2O	$BaCO_3$	$BaCO_3$	$NaNO_3$	H_2O	CO_2
D_2	CH_3OH	CH_3OH	$NaNO_2$	O_2	$BaCO_3$
$LiBD_4$	CH_3I	CH_3I	NH_4Cl	CO_2	H_2O
B_2D_6	$NaCN$	KCN	N_2H_4	CO	Me_2SO
CD_3OD		$^*CH_3CO_2Na$	$CuCN$		
C_6D_6		$CH_3{}^*CO_2Na$	$(NH_2)_2CO$		
CD_3I		C (amorphous)			
$(CD_3)_2CO$		CH_2O			

In addition, an increasing number of compounds, particularly those of bio-organic interest, are becoming available with isotopic labels.

For the manipulation of labelled reagents in synthesis, the following tips may be used to supplement the material in *Hints on the Synthesis of Organic Compounds*:

- Experience shows that if a labelled compound is commercially available, a good practical synthesis is often *not* found in the literature.
- Many labelled compounds, especially those containing ^2H, can lose much of their isotopic enrichment by exchange reactions. Syntheses must be devised to avoid exchange.
- The limiting factor in most syntheses is the cost of the labelled starting material. Design syntheses to maximise incorporation of the isotopic label, preferably by insertion of the labelled fragment as late as possible in the synthetic route.
- Compounds containing ^2H, ^{13}C, ^{14}N and ^{18}O can be handled in ordinary glass apparatus.

The preparation of isotopically labelled compounds requires special measures to be taken in order to satisfy the above considerations:

- First check all preparations using *unlabelled* material, on the same scale, at the same place and using the same apparatus.
- In some cases (and *always* in preparations using radioactive labels) it is advantageous to use a 'sandwich technique': Start the reaction with *ca.* 10 mole % of unlabelled reactant. After a suitable time (dictated by the rate at which the reaction proceeds), add the labelled sample (up to *ca.* 50 mole %) and allow the reaction to proceed for the usual period. *Then* add sufficient unlabelled material to complete the transformation. *Always* test the 'sandwich' reaction in *advance* using only unlabelled materials, but following *exactly* the conditions to be employed. The time is well spent!

- If ^2H is to be incorporated at a high level, it is usually worth using specific reactions. Exchange reactions with D_2O at equilibrium produce significantly lower deuterium levels than were present in the original D_2O. Thus, methanol-d_1 may be prepared by the hydrolysis of dimethyl carbonate with D_2O.

Special Hints

Deuteriated Compounds
In many compounds deuterium may be readily exchanged by protons. In such cases it is essential to protect the materials from any contact with moisture (see *Some Tips on Handling Water- or Air-Sensitive Substances*).

Deuterium Oxide
This is very hygroscopic: a single transfer from one vessel to another can reduce the deuterium content by 0.015%. It is best to store it under a small positive pressure of argon.

Deuterium Gas
Used for deuterium incorporation via the reduction of double and triple bonds. Note the following points:
- Use proton-free catalysts: wash with D_2O and dry in an atmosphere of D_2 gas.
- Use no protic solvents for the reduction for fear that the catalyst may also promote deuterium–proton exchange!
- The gas reserve and pressure equilibration vessels should be filled with liquid paraffin, not water.

Sodium Borodeuteride ($NaBD_4$)
Protons are exchanged for deuterium on contact with water; atmospheric moisture must be rigorously excluded. It is an advantage to add *ca.* 0.5% deuterium oxide to sodium borodeuteride for reductions in aprotic solvents if high deuterium incorporation is required: Heat the borodeuteride to 120°C and then allow it to cool in a dry desiccator in which a small vial containing the requisite amount of D_2O has been placed. The optimal reaction conditions, which lead to the utilisation of all four deuterium atoms, should be determined empirically using unlabelled material.

Lithium Borodeuteride ($LiBD_4$)
Intrinsically more reactive and more sensitive to moisture than $NaBD_4$ and should, therefore, always be transferred and reacted under protective gas. For many purposes it can be prepared *in situ* by the reaction of sodium borodeuteride and lithium chloride in dry diglyme.

Lithium Aluminium Deuteride ($LiAlD_4$)
Very sensitive to moisture. If loss of isotope is unacceptable, it is essential to work in a dry box.

Deuterodiborane (B_2D_6)
Best prepared *in situ* from dry sodium borodeuteride and boron trifluoride-etherate.

Deuterotrifluoroacetic Acid (CF₃COOD)

Deuterotrifluoroacetic Acid (CF_3COOD)

Mainly used to catalyse proton-deuteron exchange, especially in NMR spectroscopy. It is very poisonous and hygroscopic and should only be handled in serum ampoules from which it can be withdrawn using a dry syringe.

^{13}C *compounds*

The majority of ^{13}C-labelled starting materials (see list above) are one-carbon units. Unlike deuterium, ^{13}C exchange is often negligible in practice. However, one needs to avoid decarboxylations, decarbonylations and other cleavage processes, which may reduce ^{13}C incorporation.

12.5.2 Synthesis with Radioisotopes (^{14}C, ^{3}H)

Except for a few special, highly enriched tritium compounds, *e.g.* ^{3}H₂ gas, even so-called 'high-level' radioactively labelled substances contain only a small proportion of labelled molecules diluted with large quantities of compound containing only the stable isotopes. This is especially important in the case of tritium (very high isotope effect: $k_H/k_T = 7:1 - 14:1$) – any step in which complete reaction with the labelled atom does not occur will result in a major loss of label; *e.g.*, transfer of ^{3}H (cation) to a carbanion using [^{3}H]H₂O results in preferential reaction with ^{1}H – in such cases use [^{3}H]CF₃CO₂H (no enolisable H) prepared by the reaction of equimolar amounts of [^{3}H]H₂O and (CF₃CO)₂O, to ensure complete transfer of ^{3}H. By contrast, H transfer by equilibration (e.g. tritiation of an aromatic ring by prolonged reaction under acidic conditions) may use [^{3}H] H₂O efficiently.

Before embarking on reactions with radioactive substances with activities above the legal limits, make sure you know the relevant rules and regulations and that you have been properly authorised [20,21]. Radioactive compounds should only be used in specially equipped radiochemical laboratories. Whenever such compounds are being manipulated, wear gloves, laboratory coats and safety spectacles. One often needs techniques quite different from those used in conventional chemistry. Substances are frequently transferred using vacuum lines. All reactions must be carried out at all times on safety trays, so that there is no danger of the laboratory being contaminated by radioactivity. There are strict regulations governing the disposal of radioactive wastes.

For further details, consult the works listed at the end of this chapter [22].

Bibliography

[1] O. F. Shriver and M. A. Drezdzon, *The Manipulation of Air-sensitive Compounds*, 2nd edition, John Wiley & Sons, 1986.

[2] *Organic Syntheses*, 59, 124.

[3] (a) Atmos Bag, Aldrich Chemical Company; (b) I2R-Polyethylene Glove bags, Alpha Division of Ventron Company.

[4] K. Hafner, A. Stephen and C. Bernhard, *Liebigs Ann. Chem.*, 1961, 54, 650.

[5] Fluka, Aldrich and other suppliers.

[6] Fluka, Aldrich and other suppliers.

[7] D. Seebach and A. Hidber, *Chimia*, 1983, 37, 449.

[8] A. Perry and A. Weissberger. *Techniques of Chemistry*, 3rd edition, Vol. XIII, Chapter 11.5, John Wiley & Sons, Ltd, 1979.

[9] O. F. Shriver and M. A. Drezdzon. *The Manipulation of Air-sensitive Compounds*, 2nd edition, Chapter 5.4, John Wiley & Sons, 1986.

[10] E. L. Eliel and S. H. Wilen. *Stereochemistry of Organic Compounds*, John Wiley & Sons, New York, 1994.

[11] H. B. Kagan. *Asymmetric Synthesis*, Thieme Medical Publications, 2003.

[12] L. A. Paquette, ed., *Handbook of Reagents for Organic Synthesis: Chiral Reagents for Asymmetric Synthesis*, John Wiley & Sons, 2003.

[13] D. Enders and K.-E. Jaeger, eds., *Asymmetric Synthesis with Chemical and Biological Methods*, Wiley-VCH, 2006.

[14] F. Toda. *Enantiomer Separation: Fundamentals and Practical Methods*, Springer Verlag, 2005.

[15] T. E. Beesley and R. P. W. Scott. *Chiral Chromatography*, John Wiley & Sons, Inc., New York, 1999.

[16] D. Seebach, U. Gysel, and J. N. Kinkel. *Chimia*, 1991, 45, 114.

[17] J. D. Morrison, ed., (a) *Asymmetric Synthesis*, Vol. 1, *"Analytical Methods"*, (b) vols. 2 and 3, *"Stereodifferentiating Addition Reactions"*, (c) vol. 4, *"The Chiral Pool and Chiral Sulfur, Nitrogen and Silicon Centers"*, (d) Vol 5, *"Chiral Catalysis"*, Academic Press, 1983–1985.

[18] E. N Jacobsen, A. Pfaltz, and H. Yamamoto. *Comprehensive Asymmetric Catalysis*, Vols. I–III + Suppl. 1 and 2, Springer Verlag, 2004.

[19] S. V. Malhotra, ed., *Methodologies in Asymmetric Catalysis*, American Chemical Society, 2004.

[20] A U.S. guide for users of radioactive substances is found under http://www.niehs.nih.gov/odhsb/radhyg/radguide/home.htm

[21] A UK guide is available under http://www.hse.gov.uk/radiation/ionising/publications.htm

[22] H. Ross, J. E. Noakes, and J. D. Spaulding. *Liquid Scintillation Counting and Organic Scintillators*, CRC Press, 1991.

[23] W. Carruthers and I. Coldham. *Modern Methods of Organic Synthesis*, 4th edition, Cambridge University Press, 2004.

[24] L. Kurti and B. Czako. *Strategic Applications of Named Reactions in Organic Synthesis*, Academic Press, 2005.

[25] M. B. Smith and J. March. *March's Advanced Organic Chemistry, Reactions, Mechanisms, and Structure*, 5th edition, John Wiley & Sons, 2001.

[26] T. W. Greene and P. G. M. Wuts. *Protective Groups in Organic Synthesis*, 3rd edition, John Wiley & Sons, 1999.

[27] T-L. Ho. *Tactics of Organic Synthesis*, John Wiley & Sons, 1994.

[28] L. F. Armarego and C. Chai. *Purification of Laboratory Chemicals*, Butterworth-Heinemann, Oxford, 2003.

[29] Reagents for Organic Synthesis, vols. 1–22 (Vol. 22: 2004; Coll. Index for vols. 1–22: 2005) L. F. Fieser and M. Fieser, John Wiley & Sons, 1967–2005.

[30] J. R. Heys and D. G. Mellilo. *Synthesis and Applications of Isotopically Labelled Compounds 1997*, John Wiley & Sons, Ltd., Chichester, 1998.

[31] A. F. Thomas. *Deuterium Labelling in Organic Chemistry*, Meredith Corporation, 1971.

[32] D. G. Oh. *Synthesis with Stable Isotopes of Carbon, Nitrogen and Oxygen*, John Wiley & Sons, 1981.

Chapter 13

Disposal and Destruction of Dangerous Materials

Waste disposal is a problem that urgently requires a comprehensive solution. Organic research laboratories will always produce waste, so it is essential that great care is exercised in dealing with it. Because such waste often contains substances whose danger can easily be underestimated, particular care is necessary in dealing

Practical Organic Synthesis: A Student's Guide   R. Keese, M.P. Brändle and T.P. Toube
© 2006 John Wiley & Sons, Ltd.

with it. It is essential that the disposal of side-products and waste is considered when a reaction is being planned. In particular, attention should be paid to the possibility of recovering solvents. Reactions should be carried out on the smallest appropriate scale to minimise the amount of waste produced; ideally one should order no more of the required chemicals than will actually be needed. Unused chemicals should definitely be used for further applications as long as their purity has not been compromised.

Apart from harmless waste which can be incinerated without danger, all other laboratory chemicals, including solvents, should be treated as requiring special treatment. Some solvents and chemicals can of course be disposed of as normal waste, but dangerous materials must definitely be deactivated first.

The following recommendations comprise reasonable and valid procedures, but may not conform to local legal requirements. However, they all convert hazardous waste into material suitable for safe disposal.

13.1 CLASSIFICATION OF WASTE MATERIAL

In the legal sense, dangerous materials can be considered as any materials which are poisonous, corrosive, pyrophorically oxidising, flammable, explosive or strongly odorous. They include alkali metals, hydrides, organometallic compounds, poisonous gases, cyanides, haloacids, diazo and nitro compounds, peroxides and peroxyacids, certain phosphorus compounds, chlorosulfonic and concentrated sulfuric acids, mercury and its compounds, heavy metal salts and other laboratory chemicals. Dangerous materials should be rendered safe on the spot, in the laboratory, as soon as possible.

Once one has located the correct procedure for deactivating the particular class of substance, one should follow the following precautions:

- Take great care and do not rush.
- Do not use the sink drain.
- Do not work alone.
- Make sure you are wearing safety spectacles and, if necessary, rubber gloves.
- Work in an empty fume hood.
- Add substances slowly and in the correct order.
- Use an excess of the reagents to ensure complete deactivation.
- Be prepared to cool or stir the mixture, or to use an inert atmosphere (nitrogen, argon).
- Any material which cannot be treated on the spot should be sealed and labelled with the contents, date and your name, before submitting it for specialist disposal.
- If in doubt, contact the safety authorities.

13.2 ENVIRONMENTALLY ACCEPTABLE SUBSTANCES FOR TREATING DANGEROUS MATERIALS IN THE LABORATORY

Precipitants	H_2S	hydrogen sulfide
	Na_2S	sodium sulfide
	$(NH_4)_2S$	ammonium sulfide
	NaOH	sodium hydroxide
	$Ca(OH)_2$	calcium hydroxide
	$FeCl_3$	iron(III) chloride
Neutralisers for Bases	HCl	hydrochloric acid
	H_2SO_4	sulfuric acid
	$NaHSO_4$	sodium hydrogen sulfate
	H_2NSO_3H	amidosulfonic acid
Neutralisers for Acids	NaOH	sodium hydroxide
	Na_2CO_3	sodium carbonate
	K_2CO_3	potassium carbonate
	NH_4OH	ammonium hydroxide
Oxidants	NaOCl	sodium hypochlorite
	H_2O_2	hydrogen peroxide
Reductants	$NaHSO_3$	sodium hydrogen sulfite
	$Na_2S_2O_5$	sodium pyrosulfite
	$Na_2S_2O_3$	sodium thiosulfate
	$FeSO_4$	iron(II) sulfate

13.3 SPECIAL INSTRUCTIONS FOR THE DESTRUCTION OF THE MAIN CLASSES OF DANGEROUS SUBSTANCES

Substance	Method for destruction
Acids (conc. sulfuric acid, chlorosulfonic acid, methanesulfonic acid)	Dilution with water is exothermic, so cooling is essential. Add slowly to ice-water with stirring – Do not add water to acid. Then neutralise.

Substance	Method for destruction
Acid chlorides/bromides and anhydrides	Add carefully to water, cooling if necessary. Then neutralise the resulting acid.
Aldehydes	Treat water-soluble aldehydes with excess sodium sulfite solution, checking for complete reaction with starch iodide paper. Less soluble aldehydes are mixed with water and oxidised with a 20% excess of $KMnO_4$, heating if necessary; cool and treat with 6 N H_2SO_4, then reduce the remaining permanganate with sodium sulfite solution.
Alkali metals, alkaline-earth metals	Drop potassium in small pieces into dry *t*-butanol, sodium or lithium into *iso*-propanol. Heat under reflux as oxide crusts can retard the reaction. When reaction is complete, add methanol carefully and then dilute with water. Alkaline-earth metals are treated similarly, but calcium can be dropped directly into water under an inert atmosphere. Neutralise.
Alkali metal amides and hydrides	Suspend in a suitable solvent (dioxan, tetrahydrofuran) and add *iso*-propanol or ethanol under protective inert gas until a solution is obtained. The reaction is exothermic, so cool in an ice bath. Dilute carefully with water and neutralise (the final organic products are the solvent and the alcohol).
Alkali borohydrides	Dissolve in ethanol or methanol and dilute with water. Add HCl dropwise under inert gas and allow to stand until conversion into boric acid is complete (the final organic product is the alcohol).
Alkyl azides	Add ceric ammonium nitrate, with cooling.

Substance	Method for destruction
Aluminium alkyls	For hydrides, first decompose the hydride using ethyl acetate (see lithium aluminium hydride). Aluminium alkyls can be decomposed by adding methanol; cool the exothermic reaction. When the reaction dies down, dilute with water (the final organic products are the solvent, the alcohol and the alkane).
Aluminium trihalides	see: Acid chlorides.
Azide salts	Add dropwise a solution of an acid chloride in dry acetone (5 mol/l) to an aqueous solution of the azide (5 mol/l) at 10 °C, cooling the exothermic reaction. Stir for 1 h at 10–15 °C, then warm to 60 °C until N_2 evolution ceases (the final organic products are acetone and the amine).
Azocompounds, diazoesters and diazoalkyls	Add 2 N HCl dropwise to a solution in a suitable solvent, cooling the exothermic reaction. Monitor the reaction by the N_2 evolved (the final organic products are the solvent and the alkyl halide).
Calcium hydride	see: Alkali metal amides and hydrides.
Carbon disulfide	Mix with water and oxidise using sodium hypochlorite solution.
Catalysts (Ni, Cu, Fe, noble metals, filters or celite pads containing these catalysts)	Do not allow to suck dry as they are often pyrophoric. Cover filter residues with water and treat with HCl. Then put into sealed, labelled flasks for disposal or recovery.
Chlorosulfonic acid, concentrated sulfuric or hydrochloric acid, 'oleum'	Dilution with water is exothermic, so cooling is essential. Add slowly to ice-water with stirring – Do not add water to acid. Then neutralise.

Substance	Method for destruction
Chromium compounds and salts, chromic acid	Do not use chromic acid baths for cleaning as they damage the environment. Treat Cr(VI) solutions with H_2SO_4 or NaOH to pH 2–3, then decompose using sodium hydrogen sulphite; after 2 h adjust the pH to 8.5 with NaOH and filter off the resulting chromium(III) hydroxide. Send for disposal.
Cyano compounds (including cyanogen chloride or bromide)	Gaseous compounds should be absorbed in sodium hypochlorite. solution; others are decomposed by adding sodium hypochlorite solution dropwise. Monitor the reaction using starch iodide paper.
Dialkyl peroxides	Add to a solution containing a 10% excess of KI in glacial acetic acid, which has been acidified with conc. H_2SO_4. Heat slowly to 100 °C and maintain this temperature for 5 h. Reduce the iodine using sodium bisulphite.[*]
Dimethyl sulfate or diethyl sulfate	Add slowly to 10% NaOH at 40–50 °C, cooling to control the exothermic reaction. Neutralise (The final organic product is methanol or ethanol).
Grignard reagents	Add methanol or water carefully, dropwise under an inert gas, cooling to control the exothermic reaction (The final organic products are the solvent, methanol and the alkane).
Heavy metals and their salts.	Precipitate as insoluble compounds (carbonates, hydroxides, sulfides *etc.*), then send to specialised firms for disposal.
Hexamethylphosphortriamide (HMPA)	Reflux for 5 h with conc. H_2SO_4 (750 ml per mole of HMPA).
Hydrazine	Decompose dilute aqueous solution with NaOCl.

---

[*] Merck supply PEREX-KIT for determining peroxides.

Substance	Method for destruction
Hydrazoic acid	Absorb into aqueous NaOH or ammonia; then treat as under Azides.
Hydrofluoric acid	Neutralisation is exothermic, so cooling may be required. Absorb HF vapour in water in a Teflon or polythene vessel, then precipitate as insoluble calcium fluoride using calcium hydroxide solution (best at pH 12).
Hydroperoxides, peroxides	Add slowly to a solution of iron(II) sulfate (50% excess) at room temperature.
Lithium aluminium hydride	Suspend in dry dioxan (tetrahydrofuran, ether), add ethyl acetate dropwise (cool if necessary), then dilute with water and neutralise.
Mercaptans	Mix with water and oxidise using sodium hypochlorite solution.
Mercury	Pick up metal quickly, then to specialised firms (or recover). Commercial adsorbents[**] can be used to clear up spillages safely and rapidly (sprinkling sulfur or zinc powder is ineffective).
Mercury salts	Solutions can be precipitated as the sulfide using H_2S or ammonium/sodium sulfide. Filter off and send for disposal.
Metal carbonyls	Oxidise in tetrahydrofuran or hexane solution using sodium hypochlorite solution (at least 25% excess). Filter off the resulting hydroxide.
Nitriles	If necessary, reflux for several hours with excess ethanolic NaOH or 36% HCl.
Nitrites	Decompose in cold aqueous solution at pH 4 using sulfamic acid, then acidify the strongly basic solution with H_2SO_4. When N_2 evolution ceases, dilute with water.

[**] e.g., Mercurisorb Roth, Roth AG, Reinach.

Substance	Method for destruction
N-Nitroso compounds	Small amounts (mg) can be decomposed at room temperature using dilute HBr in acetic acid. Otherwise, dissolve in water, treat with NaOH solution and add aluminium–nickel alloy slowly, with cooling.
Organolithium compounds	Dissolve in a suitable solvent (dioxan, tetrahydrofuran) and add ethanol dropwise under protective inert gas. The reaction is exothermic, so cool in an ice bath. Dilute carefully with water and neutralise.
Osmium tetroxide	React with an alkene and then stir for several hours with a solution of sodium sulfite or sodium bicarbonate in aqueous methanol. The adduct can also be reduced using H_2S. Filter off the osmium dioxide and treat it as a hazardous waste.
Peroxyacids in solution	Reduce in solution in aqueous acid (with added ethanol, if necessary) with sodium hydrogen sulfite and/or an iron(II) salt and then neutralise (The final organic product is the alcohol).
Phenols	Decompose by adding 30% aqueous H_2O_2 dropwise to an aqueous solution at pH 5–6 containing iron(II) sulfate.
Phosgene	Absorb phosgene vapour in 25% NaOH; decompose phosgene in solution by adding 15% NaOH dropwise and then stirring for 1 h at room temperature.
Phosphate esters	Reflux for 1–2 h in ethanolic 2 N NaOH.
Phosphorus penta- and trihalides	The manufacturers specification recommends hydrolysis by adding an acid (*e.g.* 48% HBr) dropwise under inert gas, in order to avoid formation of spontaneously flammable phosphines. This procedure is only satisfactory with old, partially decomposed phosphorus halides.

Substance	Method for destruction
Phosphorus pentoxide	see under: Acid chlorides/bromides and anhydrides.
Silyl halides	Dissolve in a suitable solvent (toluene, hexane, dioxan and tetrahydrofuran) and add ethanol cautiously dropwise. The reaction is exothermic, so cool in an ice bath. Dilute with water or dilute NaOH (the final organic products are the solvent, ethanol and the alkyl siloxane).
Sulfides, mercaptans and thiophenols	Gaseous compounds should be absorbed in sodium hypochlorite solution; others are decomposed by adding sodium hypochlorite solution dropwise. Monitor the reaction using starch iodide paper.
Sulfur dioxide	Absorb in dilute NaOH.
Sulfur trioxide	Absorb in conc. H_2SO_4. Then treat as under: Acids.
Thionyl chloride, sulfuryl chloride	see: Acid chlorides.

Further information on the destruction of dangerous chemicals can be found in some suppliers catalogues and in the references below[1]–[4].

13.4 DEACTIVATION OF UNKNOWN LABORATORY CHEMICALS

Laboratory chemicals often have labels which have become illegible or may even be missing entirely. Although one cannot always assume that they are old, they should be treated with great circumspection. Particularly if they are suspected of being explosive, the safety authorities should be consulted.

Spectroscopic techniques can be used to identify unknown substances in the first place; they can also be used for unknown reaction products. As the substances may be poisonous or hazardous, they need to be deactivated before disposal. Simple preliminary tests and functional group reactions may indicate how unknown materials can be deactivated for disposal; it is particularly important to ascertain whether chlorine-containing compounds are present.

Preliminary Tests

Colour and **odour** may indicate the presence of certain classes of material: coloured (not discoloured) substances include aromatic, nitro, nitroso and azo compounds, amines and phenols.

Heating in a flame will determine whether or not the substance leaves a residue on combustion. Flame colour may indicate whether a compound is rich in oxygen (bluish flame which is scarcely luminous) or aromatic or unsaturated (luminous, often sooty flame). Solid residues can be tested for water solubility and pH. Heating the spatula about 1 cm from the sample provides an indication of the melting point, volatility and charring of the material.

The **solubility** of the unknown can be tested using the series: water – ether – dichloromethane – dilute NaOH – dilute HCl – more polar organic solvents such as methanol or acetone. Substances soluble in ether or dichloromethane may be identifiable by tlc. Further structural information may be gained from IR and NMR spectra.

Tests for halogen

Beilstein test
Heat a small ring of copper wire to glowing in a flame and then dip into the unknown substance; the presence of a halogen-containing substance is shown by a green or turquoise colour when the wire is then heated at the edge of a nonluminous Bunsen flame (although nitrogen-containing compounds may also give a green colouration).

Halide test
Aliphatic halides in aqueous nitric acid solution give a precipitate with silver nitrate (but aromatic or vinyl halides do not).

Sodium fusion test

A small piece of clean sodium is heated to glowing in a fusion tube with 5–20 mg of the substance; the residue can be tested for nitrogen (as CN^-), sulfur (as S^{2-}) and halide.

Oxidation

Decolourisation by the substance (*ca.* 0.1 g) in 2 ml of water or acetone of up to three drops of 2% aqueous permanganate solution indicates the presence of alkenes, phenols, enols, thiols, amines, aldehydes or alcohols.

13.5 WASTE DISPOSAL

Garbage
Harmless, solid residues which can be safely incinerated or dumped and which contain no special waste; it can be disposed of in sealed, labelled bags.

Special Waste
The following are collected separately:

• Nonhalogenated solvents.
• Chlorinated or halogenated solvents (including water containing dichloromethane).
• Acids: a metal container for each acid.

- Other liquid waste (such as formalin solution, cyanide baths and waste oil).
- Photographic chemicals.
- Solid waste (such as silca gel and salts).

Radioactive Material
Special procedures are required. Waste should be sealed in suitable containers and sent for specialist disposal.

In law, the one who produces a special waste is generally held to be responsible for its safe delivery to the collection point, so it is advisable to obtain a receipt. Packages containing such waste should be clearly labelled 'SPECIAL WASTE'.

Bibliography

[1] *Prudent Practices in the Laboratory: Handling and Disposal of Chemicals*, National Research Council, National Academy Press, 1995, Chapter 6.
[2] G. Lunn and E B. Sansome. *Destruction of Hazardous Chemicals in the Laboratory*, 2nd edition, John Wiley & Sons, Inc., 1994.
[3] M. A. Armour. *Hazardous Laboratory Chemicals Disposal Guide*, 3rd edition, CRC Press, 2003.
[4] *Laboratory Waste Management: A Guidebook*, American Chemical Society, 1994.
[5] The Material Safety Data Sheet (MSDS's, *cf.* Chapter 2) of any particular substance contains information about the specific disposal procedure.

Chapter 14

Purification and Drying of Solvents

Purity, as applied to a solvent, is not an absolute concept. It is impossible to prove the total absence of a particular impurity: a negative result for any chemical or physical test merely shows that the concentration of a substance is below the detection limit of the test. In practice, an organic chemist is less interested in this philosophical 'absolute purity' than in the suitability of a solvent for a particular application or reaction.

Practical Organic Synthesis: A Student's Guide   R. Keese, M.P. Brändle and T.P. Toube
© 2006 John Wiley & Sons, Ltd.

Careful purification (and drying – water is also an impurity) of solvents usually proves worthwhile. In particular, it is advisable to use freshly distilled solvents at all times. The apparatus shown is suitable for keeping freshly distilled solvents continuously available. If the lateral outlet of the dropping funnel is sealed using a serum cap, fresh solvent can be drawn off using a syringe.

In this section the main drying agents are described, and simple procedures are given for purification and drying of the most common solvents, together with their properties. Quite apart from anything discussed here, the demands on the purity of solvents

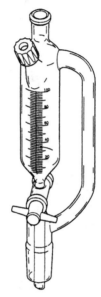

○ Figure 14.1: Pressure-equalising graduated funnel for storage of freshly distilled solvent

for particular applications should always be considered anew. These procedures can only act as guides in this matter.

Improper handling of most solvents can lead to serious health hazards. Accordingly, the exposure limit values (OES) (see Chapter 2) for each solvent, i.e. the maximum concentrations that one can encounter with safety in a laboratory or factory as supported by current information, have been given.

14.1 SOME DRYING AGENTS

See Chapter 13 for information on disposing of used drying agents.

14.1.1 Alumina

Al_2O_3

MW 101.96   Capacity maximum 10%.

(a) *Removal of Peroxides from Ethers and Hydrocarbons*
Alumina 100 g (activity I, basic) in a 15 mm diameter column will remove peroxides by adsorptive filtration from the following volumes of dried solvent: 1000 ml diethyl ether, 400 ml diisopropyl ether, 1000 ml tetralin and 100 ml dioxan. Basic alumina activity Super I is approximately twice as efficient.

The peroxides are adsorbed but not decomposed. Used alumina should therefore be disposed of carefully.

(b) *Purification of Hydrocarbons for UV Spectroscopy*
See *e.g.* 'Hexane'.

(c) *Removal of Water from Organic Solvents*
Economical only for solvents which have previously been dried and distilled. In that case 150 g alumina (activity I, basic) should suffice to remove residual moisture from 100 to 1100 ml of solvents containing less than 0.01% water. It is particularly suitable for hydrocarbons, as well as for many ethers and esters. In the process, peroxides and other polar impurities are removed.

(d) *Removal of Ethanol Used to Stabilise Chloroform*
See under 'Chloroform'.

14.1.2 Barium Oxide

BaO
MW 153.34
OES (dust) $0.5 \, mg \, m^{-3}$.

A relatively vigorous drying agent used for ethyl acetate and recommended for organic bases. Use freshly crushed material, as the powdery reagent is probably no longer active.

14.1.3 Calcium Chloride

$CaCl_2$ Residual water: 0.25 mg/l dry air
MW 110.99   Capacity: *ca.* 90%.

Cheap, slow and not particularly efficient. Takes up water to form the hexahydrate below 30 °C. Suitable for predrying hydrocarbons, alkyl halides, ethers and many esters.

It melts on taking up excess water. Drying tubes should be used only once, or tested before use to ensure that they are not blocked!

14.1.4 Calcium Hydride

CaH_2
MW 42.10

Very effective drying agent for alkanes, dimethylformamide, dioxan, dichloromethane, chloroform and pyridine. For disposal, see Chapter 13.

14.1.5 Calcium Oxide

CaO   Residual water: 0.2 mg/l dry air
MW 56.08

Cheap. Suitable for drying amines, but not for low-boiling alcohols. Heat together under reflux for an hour or two and then distil.

14.1.6 Calcium Sulfate – Semihydrate (Drierite, Sikkon)

$CaSO_4 \cdot 0.5H_2O$ Residual water: 0.004 mg/l dry air
MW 145.15   Capacity: *ca.* 7%.

Powerful drying agent, effective up to *ca.* 100 °C. Suitable for almost all organic liquids and gases.

14.1.7 Lithium Aluminium Hydride (LAH)

$LiAlH_4$
MW 37.95.

Very efficient at drying hydrocarbons and ethers. These solvents must be predried as LAH reacts vigorously with water.

Dangerous: it decomposes above 150 °C; distillation over LAH should never be allowed to go to dryness, even using an inert atmosphere! Decompose the excess reagent (see Chapter 13).

14.1.8 Magnesium

Mg
Atomic weight 24.31.

Used for making 'superdry' alcohols: see under 'Methanol'.

14.1.9 Magnesium Sulfate

$MgSO_4$ Residual water: 1.0 mg/l dry air
MW 120.37   Capacity: *ca.* 100% (formation of heptahydrate).

Suitable for drying almost all compounds, including acids and their derivatives, aldehydes and ketones. Standard medium for drying solutions.

For drying exceptionally acid-sensitive compounds use sodium sulphate or a basic drying agent.

14.1.10 Molecular Sieve

Molecular sieves are synthetic crystalline aluminium silicates [1]. After the water of crystallisation has been removed (i.e. in the active state), they have a large number of cavities associated with pores of a closely defined molecular diameter (3, 4, 5 or 10 Å, according to type). These cavities can then be occupied, but only by molecules having critical van der Waals radii less than the pore diameters. For example:

H_2	2.4 Å	CO_2	2.8 Å	NH_3	3.8 Å	C_2H_4	4.25 Å
H_2O	2.6 Å	N_2	3.0 Å	Cl_2	8.2 Å	C_2H_6	4.44 Å
O_2	2.8 Å	CO	3.2 Å	C_2H_2	2.4 Å	n-alkanes	4.89 Å

The maximum capacity of molecular sieve is about 20%. Its bulk density lies between 50 and 70 g per 100 ml.

Molecular sieve has the advantage of being capable of regeneration at will with effectively no loss of capacity. It should always be reactivated before use by heating in an oven at 300–350 °C either in vacuo or in a stream of nitrogen or argon. In the absence of a suitable drying oven the following procedure will do: mix used molecular sieve with plenty of water, filter and dry for a few hours in a drying cupboard at 150 °C, then overnight under high vacuum at 200 °C (silicon oil bath) or in a suitable oven at 400 °C. When dealing with large quantities it is advantageous to interrupt the vacuum drying after 2 hours and flush with argon to ensure that the granules dry adequately. This treatment gives molecular sieve with a residual water content less than 0.5% and with a 20% capacity to absorb water.

To determine its activity, add 10 ml of water to 10 g molecular sieve in a plastic vessel; fully active material causes a temperature rise of 35–40 °C, and this temperature rise is linearly proportional to the activity.

The (dynamic) drying of organic solvents is most efficient if the predried, distilled and peroxide-free solvent is percolated through a column of molecular sieve (dynamic drying). A 250 g molecular sieve column (25 mm diameter, 600 mm high) will dry 10 litres of the following solvents to a purity better than 0.002% residual water at 3 litres an hour: diethyl ether, diisopropyl ether, tetrahydrofuran, dioxan, benzene, toluene, cyclohexane, dichloromethane, chloroform, carbon tetrachloride, pyridine, ethyl acetate (all these with 4 Å sieve) and acetonitrile, acetone, ethanol, methanol and other alcohols (3 Å sieve). Stand the solvent over molecular sieve for several hours, repeating if necessary.

Dried solvents can be stored over molecular sieve (*ca.* 10 g/l); they may become cloudy, in which case they should be redistilled. Acetone and other ketones should not be stored over molecular sieve as they are sufficiently basic to catalyse aldol reactions.

14.1.11 Phosphorus Pentoxide

P_2O_5 Residual water: less than 0.000025 mg/l dry air
MW 141.94

Very rapid and efficient. One of the best drying agents, except that the surface becomes syrupy as water is taken up and this hinders further uptake. This problem is overcome in various patent preparations (e.g. Fluka phosphorus pentoxide drying agent – 75% P_2O_5 in an inert carrier, Merck Sicapent) that remain particulate even after 100% uptake of water.

Phosphorus pentoxide is suitable for drying saturated and aromatic hydrocarbons, anhydrides, nitriles, alkyl and aryl halides and carbon disulfide. It is not suitable for alcohols, amines, acids, aldehydes or ketones.

Care is required in the decomposition of large amounts of excess P_2O_5. A practical method is to add it in small portions to a large quantity of ice, and subsequently to neutralise with base.

14.1.12 Potassium Carbonate

K_2CO_3
M 138.21.

Suitable for preliminary drying of organic bases at room temperature.

14.1.13 Potassium Hydroxide

KOH   Residual water: 0.002 mg/l dry air
MW 56.11

Suitable for drying organic bases. Melts as it takes up water (use in a drying tower). It is much more efficient than sodium hydroxide.

Dangerous: work carefully, wear safety glasses!

14.1.14 Silica Gel

(Kieselgel) Capacity: *ca.* 35%

Use in desiccators and drying tubes. Best to use the granules containing an indicator that is blue when dry and pink when saturated with water.

14.1.15 Sodium

Na
Atomic weight 22.99.

Suitable for drying ethers, tertiary amines and saturated or aromatic hydrocarbons. Only predried solvents should be dried with sodium.

Sodium has been largely superseded as a laboratory drying agent by sodium or calcium hydrides, which are easier to use, equally efficient and simpler to decompose.

The blue solution of sodium in benzophenone can be used for drying ether, tetrahydrofuran and 1,2-dioxan, as long as reducible impurities are absent.

Excess sodium must be destroyed (see Chapter 13).

14.1.16 Sodium Hydride

NaH
MW 23.998.

Used usually in the form of a dispersion in oil for drying ethers and hydrocarbons. Sodium hydride may ignite spontaneously on contact with water and burn explosively (see Chapter 13).

14.1.17 Sodium Sulfate

Na_2SO_4 Residual water: 12 mg/l dry air
MW 162.04   Capacity: *ca.* 75%.

Useful for preliminary drying at room temperature of sensitive compounds such as acids, aldehydes, ketones and halides. The theoretical capacity (formation of decahydrate) is unattainable. Somewhat less efficient than magnesium sulfate.

14.1.18 Sulfuric Acid

H_2SO_4
MW 98.08

Suitable for inert neutral or acidic gases.
Available in 'granulated form' (25% inert carrier), with or without an indicator (*e.g.* Merck Sicacide).

14.1.19 Determination of Residual Water in Solvents

Karl Fischer titrations measure water content reliably in the range ppm to 100% [2]. Saturated and unsaturated hydrocarbons, haloalkanes and alcohols can be titrated directly or in methanol solution.

14.2 SOLVENTS*

(For destruction of dangerous solvents and of sodium, sulfuric acid, phosphorus pentoxide, *etc.* used in drying and purification, see Chapter 13).

In the laboratory one should handle solvents in a responsible manner. Contrary to a widespread misconception, solvents (even dichloromethane and chloroform) are significantly soluble in water. Some of them decompose very slowly and may thus

---

* Some of the OES figures are supplemented by the following symbols: A = may be absorbed through the skin; C = suspected/known carcinogen; M = requires biological monitoring; S = sensitiser.

contaminate water supplies. Their relatively low boiling points and significant volatility mean that they can cause appreciable air pollution.

There are national regulations and international recommendations aimed at minimising the environmental damage caused by solvents, particularly the chlorinated ones. Thus particular care is needed in handling solvents like dichloromethane and fluorotrichloromethane (Freon 113). Evaporation of the solvents is best carried out using a membrane pump with efficient condensation. If a water pump has to be used, a vacustat fitted with a shut-off valve to the pump inserted after the condenser may reduce the amount of solvent vapour carried away in the water.

Halogenated solvents need to be kept separate from others and, ideally, should be recovered. Follow local disposal procedures.

14.2.1 Acetone

CH_3COCH_3

MW 58.08	mp −94.7 °C	d_4^{20} 0.791
flash point	bp 56.3 °C	dielectric const.
(flame induced) −30 °C		(25 °C) 20.70
OES 750 ppm	odour threshold	
	200–450 ppm.	

Relatively harmless. Recommended for cleaning glassware. Can be recovered by distillation in favourable cases (discard grossly impure fractions into solvent residue can). Completely miscible with methanol, ethanol, ether, water, *etc.* No azeotrope with water. On contact with basic or acid reagents acetone forms condensation products [3].

Purification
Dry over copper sulfate or 3 Å molecular sieve. Fractional distillation over boric anhydride is also recommended. Store over 3 Å molecular sieve.

14.2.2 Acetic acid (Glacial Acetic Acid, Ethanoic Acid)

CH_3CO_2H

MW 60.053	mp 16.6 °C	d_4^{25} 1.044
	bp 117.9 °C	dielectric const. (20 °C) 6.15
OES 10 ppm		

No water azeotrope. Completely miscible with water. Extremely hygroscopic.

Purification
For many purposes it is sufficient to purify by crystallisation. Cool in an ice bath until it solidifies and decant the mother liquor when about 80% of the acid has crystallised; repeat twice. Can be dried over phosphorus pentoxide (but some acetic anhydride may be formed).

14.2.3 Acetic Anhydride (Ethanoic Acid Anhydride)

$(CH_3CO)_2O$
MW 102.091   mp $-73.1\,^{\circ}C$ d_4^{15} 1.087
                  bp $140.0\,^{\circ}C$ dielectric const. $(19\,^{\circ}C)$ 20.7.
OES 5 ppm! (Poisons lungs, irritates mucous membranes).
Hydrolyses very slowly in water at $0\,^{\circ}C$ and pH 7.

Purification
Stand for a few hours over phosphorus pentoxide (100 g/l), decant, stand over the same amount of anhydrous potassium carbonate for a few hours, filter and distil in the absence of moisture at 100 Torr. It can also be purified by careful fractional distillation using an efficient column.

14.2.4 Acetonitrile

CH_3CN
MW 41.053   mp $-48.8\,^{\circ}C$ d_4^{25} 0.777
                  bp $81.6\,^{\circ}C$ dielectric const. $(20\,^{\circ}C)$ 37.50
OES 40 ppm A
Water azeotrope: $76.5\,^{\circ}C$ (83.7% acetonitrile).

Completely miscible with ethanol, ether and water; slightly soluble in hydrocarbons.

Purification
Predry over calcium chloride or potassium carbonate. Distil over 1% (by weight) of phosphorus pentoxide, then from *ca.* 5% (by weight) anhydrous potassium carbonate. Distillation over calcium hydride is also effective.

14.2.5 Ammonia

NH_3
MW 17.03   mp $-77.7\,^{\circ}C$ d (at bp) 0.682
                  bp $-33.3\,^{\circ}C$ dielectric const. $(-33\,^{\circ}C)$ 22.4
OES 25 ppm

Purification
Treat condensed ammonia with sodium (2–5 g/l), allow the blue solution to reflux for 15 min, then recondense.

14.2.6 Aniline (Aminobenzene)

$C_6H_5NH_2$
MW 93.129   mp $-5.98\,^{\circ}C$ d_4^{25} 1.01750
                  bp $184.40\,^{\circ}C$ dielectric const. $(20\,^{\circ}C)$ 6.89

OES 5 ppm – also by skin adsorption, so avoid skin contact entirely.

Purification
Fractional distillation under water-pump vacuum, if necessary after treatment with barium oxide.

14.2.7 Anisole (Phenyl Methyl Ether)

$C_6H_5OCH_3$
MW 108.15   mp $-37.5\,°C$ d_4^{20} 0.99
                         bp 153.8 °C   dielectric const. (25 °C) 4.30
Solubility in water: 10.3 g/l at 25 °C.

Purification
Fractional distillation under water-pump vacuum, or dry over calcium chloride and then distil from sodium.

14.2.8 Benzene

C_6H_6
MW 78.115   mp 5.5 °C d_4^{25} 0.874
                         bp 80.1 °C   dielectric const. (25 °C) 2.28
OES 5 ppm A,C,M. Handle only in a fume-hood. Whenever possible use (less dangerous) toluene or xylene instead.

Water azeotrope: 69.25 °C (91.17%, benzene). At 20 °C it is saturated by 0.06% water. Solubility in water: 1.8 g/l at 20 °C.

Purification
(a) Distillation, discarding 10% forerun. (b) Shake repeatedly with portions (5% by volume) of conc. sulfuric acid, wash well with water until neutral, dry over calcium sulfate (100 g/l) decant, distil from sodium hydride dispersion (0.5 g/l) or calcium hydride. If this last method has to be used for toluene or xylene, prolonged contact with sulfuric acid should be avoided as sulfonation of these latter hydrocarbons is facile.

14.2.9 Butan-l-ol

$CH_3CH_2CH_2CH_2OH$
MW 74.124   mp $-88.6\,°C$ d_4^{25} 0.806
                         bp 117.7 °C   dielectric const. (25 °C) 17.51
OES 50 ppm A
Water azeotrope: 92.7 °C (57.5% butanol). At 25 °C, 7.45% soluble in water, and dissolves 20.5% water itself.

Purification
Reflux over sodium/dibutyl phthalate and distil (*cf.* 'ethanol') [3].

14.2.10 Butan-2-ol

$CH_3CH_2CH(OH)CH_3$
MW 74.124   mp −114.7 °C d_4^{25} 0.803
                   bp 99.9 °C   dielectric const. (25 °C) 16.56
OES 100 ppm
Water azeotrope: 87.0 °C (73.2% butanol). At 20 °C, 12.5% soluble in water; at 25 °C
it dissolves 44.1% water.

Purification
(a) Distil over calcium hydride or (b) reflux over sodium, add dibut-2-yl succinate and
distil [3].

14.2.11 *tert*-Butanol (2-Methylpropan-2-ol)

$(CH_3)_3COH$
MW 74.124   mp 25.8 °C d_4^{30} 0.776
                   bp 82.4 °C   dielectric const. (25 °C) 10.9
OES 100 ppm
Water azeotrope 79.9 °C (88.24% butanol). Ternary azeotrope with water and benzene 67 °C. Completely miscible with water, ethanol and ether. As it is solid at room
temperature, it will solidify in a condenser: use a circulating pump and water from a
thermostatic bath at *ca.* 30 °C.

Purification
(a) Reflux over calcium oxide (3% by weight) and then distil. (b) Predry over calcium
hydride or sodium (1%) and then distil.

14.2.12 *iso*-Butanol (2-Methylpropan-l-ol)

$(CH_3)_2CHCH_2OH$
MW 74.124   mp −108 °C d_4^{25} 0.798
                   bp 108.4 °C   dielectric const. (25 °C) 17.93
OES 50 ppm
Water azeotrope 89.8 °C (67% butanol). At 25 °C, 10% soluble in water, and dissolves
16.9% water itself.

Purification
(a) Reflux over calcium oxide and then distil using an efficient column. (b) Predry and
then distil over calcium hydride.

14.2.13 Butan-2-one (Methyl Ethyl Ketone, MEK)

$CH_3CH_2COCH_3$
MW 72.108   mp −86.69 °C d_4^{25} 0.805
                   bp 79.6 °C   dielectric const. (20 °C) 18.51
OES 200 ppm

Water azeotrope: 73.41 °C (88.73% ketone). At 20 °C, 24% soluble in water, and dissolves 10% water itself.

Purification
see 'Acetone'.

14.2.14 *t*-Butyl Methyl Ether

$(CH_3)_3COCH_3$
MW 88.15   mp −108.6 °C d_4^{25} 0.735
                      bp 55.2 °C.
Water azeotrope 52.6 °C (4% water). Ternary azeotrope with methanol/water. Solubility in water: 48 g/l at 20 °C. Stable in neutral or alkaline solution, but decomposed by mineral acids. Minimal autoxidation in storage.

Purification
see 'Diethyl Ether'.

14.2.15 Carbon Disulfide

CS_2
MW 76.139   mp −111.5 °C d_4^{25} 1.270; d_4^{25} 1.248
                     bp 46.2 °C   dielectric const. (20 °C) 2.64
OES 10 ppm A
It is extremely flammable. Mixtures of carbon disulfide and air have flash points (flame induced) of about −30 °C; ignition temperature (at a hot surface) is *ca.* 100 °C (steam bath!).

Water azeotrope: 42.6 °C (97.2% CS_2). At 20 °C, it is saturated by less than 0.005% water and dissolves 0.3% in water itself.

Purification
Distillation over a little phosphorus pentoxide (*ca.* 10 g/l).

14.2.16 Carbon Tetrachloride

CCl_4
MW 153.823   mp −22.95 °C d_4^{25} 1.584
                     bp 76.75 °C   dielectric const. (20 °C) 2.24
OES 5 ppm A,C.
Chronic exposure causes liver damage. Do not inhale vapour. Handle in a fume hood. Avoid whenever possible.
Water azeotrope: 66 °C (95.9% carbon tetrachloride). At 24 °C it is saturated by 0.010% water and itself dissolves in water 0.8 g/l. Carbon tetrachloride is not flammable.

Purification
Reflux over phosphorus pentoxide (*ca.* 5 g/l) for 30 min and then distil. For spectroscopic use, this distillate should be filtered through alumina (basic, activity 1) (see 'Chloroform').

14.2.17 Chlorobenzene

C_6H_5Cl

MW 112.560   mp $-45.5\,°C$ d_4^{20} 1.106

                 bp $131.7\,°C$ dielectric const. ($25\,°C$) 5.62

OES 50 ppm M

Water azeotrope $90.2\,°C$ (71.6% chlorobenzene). At $25\,°C$ it is saturated by 0.0327% water.

Purification

Shake repeatedly with portions (*ca.* 5 vol. %) of conc. H_2SO_4. Wash until neutral, dry (calcium chloride), decant and distil over P_2O_5 (5 g/l). If required free of all traces of acid, pass through an alumina (basic, activity I) column (100 g/l) (cf. 'chloroform'). Drying agent: molecular sieve 4 Å.

14.2.18 Chloroform

$CHCl_3$

MW 119.378   mp $-63.5\,°C$ d_4^{25} 1.480

                 bp $61.2\,°C$ dielectric const. ($20\,°C$) 4.81

OES 10 ppm

Strongly suspected carcinogen. It is anaesthetic and can cause liver damage. Exceptionally damaging to eyes (*Caution*: using in syringes e.g., for IR spectroscopy).

Water azeotrope: $56.12\,°C$ (97.8% chloroform). Solubility in water: 8.0 g/l at $20\,°C$. Not flammable.

In light it reacts with oxygen to form phosgene ($COCl_2$), chlorine, hydrogen chloride etc.; commercial chloroform is therefore stabilised by the addition of *ca.* 1% ethanol.

Ethanol-free Chloroform (e.g., *for IR spectroscopy*)

Adsorptive filtration through alumina (basic, activity 1); a 50 g column will give *ca.* 70 ml chloroform containing less than 0.005% ethanol and free of traces of water or acid. The first 25 ml should be collected and put onto the column again as the forerun tends to contain excess water; the heat of adsorption liberated when the column begins operation reduces its efficiency and it only reaches its full drying power when it has cooled. Stabiliser-free chloroform can be kept in the refrigerator in a dark bottle, but not longer than 2 weeks [4].

Purification

Distillation over phosphorus pentoxide (*ca.* 5 g/l).

Danger: chloroform may react explosively with strong bases or alkali metals!

14.2.19 Cyclohexane

C_6H_{12}

MW 84.162   mp $6.5\,°C$ d_4^{25} 0.774

                bp $80.7\,°C$ dielectric const. ($20\,°C$) 2.023

OES 300 ppm
Water azeotrope: 68.95 °C (91% cyclohexane), suitable for removal of water in esterification with ethanol.

Purification
see 'Hexane'.

14.2.20 Decalin (Decahydronaphthalene, bicyclo[4,4,0]decane) – mixture of isomers

$C_{10}H_{18}$
MW 138.255   mp −40 °C d_4^{25} 0.879
                    bp 189–191 °C   dielectric const. (25 °C) 2.15
OES not established. May cause eczema [5].

Purification
(a) Heat under reflux with sodium wire for a few hours in an inert atmosphere (nitrogen, argon) and then distil. (b) Filter through alumina (basic, activity I, *ca.* 100 g/l). Forms peroxides with oxygen in light, so store under nitrogen or argon. Can be dried over molecular sieve 4 Å but peroxides are not then removed. (cf. 'Diethyl Ether').

14.2.21 1,2-Dichloroethane (Ethylene Chloride)

$ClCH_2CH_2Cl$
MW 98.96   mp −35.6 °C d_4^{25} 1.246
                  bp 83.5 °C   dielectric const. (25 °C) 10.66
OES 10 ppm A
Water azeotrope 72 °C (91.8% dichloroethane). Azeotropes with methanol, ethanol, propan-2-ol and carbon tetrachloride.

At 20 °C it is saturated by 0.15% water and itself dissolves in water 8 g/l.

Purification
(a) Distil over phosphorus pentoxide (5 g/l) and then filter through alumina (basic, activity 1, 50 g/l). (b) Distil over calcium hydride. If all traces of olefin need to be removed, before distilling shake several times with portions of concentrated sulfuric acid (5% by volume), wash until neutral and predry over magnesium sulfate.

14.2.22 Dichloromethane (DCM, Methylene Chloride)

CH_2Cl_2
MW 84.933   mp −95.1 °C d_4^{25} 1.317
                   bp 40.2 °C   dielectric const. (25 °C) 8.93
OES 100 ppm M (much less poisonous than chloroform, OES 50 ppm, or carbon tetrachloride, OES 10 ppm).

Water azeotrope 38.1 °C (98.5% dichloromethane). At 25 °C saturated by 1.30% water and itself dissolves in water 0.17 g/l.

Purification
(a) Reflux for 30 min over phosphorus pentoxide (5 g/l) and then distil. Traces of acid in the distilled solvent can be removed by filtration through alumina (basic, activity 1, 50 g/l). (b) Distil over calcium hydride. For distillation use a membrane pump with an efficient condenser.

Danger: Dichloromethane reacts explosively with alkali metals or strong bases!

14.2.23 Diethyl Carbonate

$(CH_3CH_2O)_2CO$
MW 118.134   mp −43.0 °C d_4^{25} 0.969
                bp 126.8 °C   dielectric const. (20 °C) 2.820
OES not known

14.2.24 Diethylene Glycol Dimethyl Ether (Diglyme)

$CH_3OCH_2CH_2OCH_2CH_2OCH_3$
MW 134.177   mp −64 °C d_4^{25} 0.944
                bp 150 °C(decomposition)   dielectric const. (20 °C) 7
OES not known. It is poisonous.
Hygroscopic. Totally miscible with water and most of the common solvents. May contain 2,6-di-*tert*.-butyl-4-methylphenol as a stabiliser.

Purification
(a) Fractional distillation under reduced pressure (water-pump, bp 62–63 °C/15 Torr). (b) Treat with several batches of fresh sodium wire until no further reaction is detected, decant and fractionate in inert atmosphere. (c) Stand for several hours over calcium hydride, then reduced pressure fractional distillation. Lithium aluminium hydride is prone to react explosively with diglyme at 180–200 °C.

14.2.25 Diethyl Ether (Ether)

$CH_3CH_2OCH_2CH_3$
MW 74.124   mp −116 °C d_4^{25} 0.7076
                bp 34.6 °C   dielectric const. (20 °C) 4.335
OES 400 ppm
Ether vapour is highly flammable with a flash point (flame induced) of −45 °C and an ignition temperature (at a hot surface) of 180 °C; it is heavier than air and it will creep along a bench top for a considerable distance.

At 25 °C, 6% soluble in water, and dissolves 1.468% water itself.

Commercial ether contains variable amounts of water, ethanol and peroxides.

Purification
Add sodium hydride (55–60% dispersion in oil; *ca* 1 g/l) slowly to the ether, reflux for 30 min and then distil. Ether suitable for almost all applications can be prepared by a second distillation, this time in an inert atmosphere (nitrogen, argon) from lithium aluminium hydride (0.5 g/l).

In the presence of light, ether combines with oxygen to form peroxides which are concentrated on distillation and can explode violently. Accordingly, unpurified ether should *never* be taken to dryness. The presence of peroxides can be detected as follows: shake 3 ml ether with a solution of potassium iodide (200 mg) in hydrochloric acid (1 M, 10 cm³) – a brown colour (liberated iodine) indicates that peroxides are present. Ether should be stored in the absence of light. Bottles should always be used within a few days of being opened (avoid a large air volume above the ether).

Ether can also be purified by distillation from sodium and filtration through alumina (basic, activity 1) (100 g/l). The spent alumina should not be heated, but disposed of carefully, as the peroxides are adsorbed, not destroyed. Alternatively, dry over molecular sieve 4 Å (which does not remove peroxides – a mixture of molecular sieve and alumina may be used if necessary).

14.2.26 1,2-Dimethoxyethane (Monoglyme, Ethylene Glycol Dimethyl Ether)

CH₃ OCH₂ CH₂ OCH₃

MW 90.12	mp −58 °C	d_4^{20} 0.867
	bp 93 °C	dielectric const. (25 °C) 7.20

OES: not known, but it is moderately irritant to skin and mucous membranes.

Water azeotrope (89.9% monoglyme). Completely miscible with water and many organic solvents. May contain 2,6-di-*tert*-butyl-4-methylphenol as a stabiliser.

Purification
(a) Treat with several batches of fresh sodium wire until no further reaction is detected and then distil from sodium/benzophenone once the blue colour persists. (b) Predry over calcium hydride and then distil from sodium/benzophenone once the blue colour persists.

14.2.27 *N,N*-Dimethylformamide (DMF, Formdimethylamide)

(CH₃)₂NCHO

MW 73.095	mp −60.4 °C	d_4^{25} 0.944
	bp 153.0 °C	dielectric const. (25 °C) 36.71

OES 10 ppm A, M
No water azeotrope.

At 150 °C it slowly decomposes to give dimethylamine and carbon rnonoxide. Decomposition is base-catalysed and will proceed at room temperature, e.g. in the presence of sodium hydroxide. DMF is completely miscible with water and many organic solvents. It is a good solvent for salts.

Purification
(a) Mix DMF (250 g), cyclohexane (30 g) and water (12 g). Distil at 140 °C to remove impurities together with the water and cyclohexane. Cool (exclude moisture) and then distil under reduced pressure (water-pump). Store in the dark. (b) Stand over P_2O_5 for 24 h, then distil. Store over 3 Å molecular sieve. (c) Can also be distilled over calcium hydride [5].

14.2.28 Dimethylsulfoxide (DMSO)

$(CH_3)_2SO$
MW 78.134   mp 18.5 °C d_4^{25} 1.096
                   bp 189 °C   dielectric const. (25 °C) 46.68
OES 50 ppm. Readily adsorbed through the skin, and it carries many substances through with it. Avoid *all* contact with the skin. No water azeotrope.

Purification
Distil under reduced pressure (water-pump, bp 75 °C/12 Torr). Dry repeatedly over 3 Å molecular sieve.

14.2.29 1,4-Dioxan (Dioxan, p-Dioxan)

$C_4H_8O_2$
MW 88.107   mp 11.8 °C d_4^{25} 1.028
                   bp 101.3 °C   dielectric const. (25 °C) 2.21
OES 25 ppm A. Possibly carcinogenic.

Water azeotrope: 87.82 °C (82% dioxan). Completely miscible with water. Forms peroxides (for detection, see 'Diethyl ether').

Purification
Predry over solid potassium hydroxide, decant and distil over calcium hydride or sodium/benzophenone. Store in inert atmosphere (nitrogen, argon). Can be dried over 3 Å molecular sieve. Molecular sieve will not remove peroxides.

14.2.30 Diisopropyl Ether

$(CH_3)_2CHOCH(CH_3)_2$
MW 102.187   mp −85.5 °C d_4^{25} 0.718
                   bp 68.3 °C   dielectric const. (25 °C) 3.88
OES 250 ppm

Water azeotrope: 62.2 °C (95.5% ether). At 20 °C it is saturated by 0.58% water and dissolves 0.9% water itself.

Purification
see 'Diethyl ether'.
Danger: Forms peroxides with oxygen very readily!

14.2.31 Ethanol (Ethyl Alcohol)

CH_3CH_2OH
MW 46.080   mp −114.1 °C d_4^{25} 0.785
              bp 78.4 °C   dielectric const. (25 °C) 24.77
OES 1000 ppm
Water azeotrope: 78.14 °C (96% ethanol). Ternary azeotrope with water/cyclohexane 62.1 °C (17% ethanol, 76% cyclohexane and 7% water). This ternary azeotrope, which does not separate into two phases on cooling, can be used to dry ethanol or for esterification.

Completely miscible with benzene, chloroform, ether, acetone and water.

Purification
Treat commercial absolute alcohol (which may contain 0.01% water or more) with sodium (*ca.* 7 g/l) in the absence of moisture. When ethoxide formation is complete, add *ca.* 30 g diethyl phthalate, reflux for 2 h and distil.

A less efficient Purification: Reflux SVR (96%) ethanol over *ca.* 25% calcium oxide for 15 h and then distil.

Further drying: Magnesium (see under 'Methanol') or calcium hydride.

14.2.32 Ethyl Acetate

$CH_3CO_2CH_2CH_3$
MW 88.107   mp −83.6 °C d_4^{25} 0.895
              bp 77.1 °C   dielectric const. (25 °C) 6.02
OES 400 ppm
Water azeotrope: 70.38 °C (91.53% ester). At 25 °C 8.08% soluble in water and it takes up 2.94% water itself.

Purification
(a) Reflux for 2 h over barium oxide (*ca.* 5 g/l) and then distil. (b) Dry over phosphorus pentoxide or 4 Å molecular sieve and then distil.

14.2.33 Ethylene Chloride

see '1,2-Dichloroethane'.

14.2.34 Ethylene Glycol (Ethane-1,2-diol)

$HOCH_2CH_2OH$
MW 62.029   mp $-13\,°C$ (glass) d_4^{25} 1.110
                bp 197.6 °C   dielectric const. (25 °C) 37.7
OES 10 ppm A. It is poisonous (cf. methanol) unlike glycerine.
No water azeotrope.

Purification
Azeotropic distillation with benzene, then reduced pressure distillation. Can be dried over magnesium (see 'Methanol'). Inclined to bump in boiling.

14.2.35 Ethylene Glycol Dimethyl Ether

see '1,2-Dimethoxyethane'.

14.2.36 Ethylene Glycol Monoethyl Ether (Ethyl Cellosolve, 2-Ethoxyethanol)

$CH_3\ CH_2\ OCH_2\ CH_2\ OH$
MW 90.123   mp below $-90\,°C$ d_4^{25} 0.925
                bp 135.6 °C   dielectric const. (24 °C) 29.6
OES 5 ppm A

Purification
Distillation.

14.2.37 Ethylene Glycol Monomethyl Ether (Methyl Cellosolve, 2-Methoxyethanol)

$CH_3OCH_2CH_2OH$
MW 76.096   mp $-85.1\,°C$ d_4^{25} 0.960
                bp 124.6 °C   dielectric const. (25 °C) 16.23
OES 5 ppm A

Purification
Distillation.

14.2.38 Ethyl Formate

$HCO_2CH_2CH_3$
MW 74.080   mp $-79.4\,°C$ d_4^{25} 0.924
                bp 54.2 °C   dielectric const. (25 °C) 7.16
OES 100 ppm
Water azeotrope: 52.6 °C (95% ester). At 20 °C it dissolves 17% water. Solubility in water: 118 g/l at 20 °C.

Purification
Dry over anhydrous magnesium sulfate, decant and then distil over phosphorus pentoxide (*ca.* 10 g/l).

Further drying agents: anhydrous sodium sulfate or potassium carbonate (not calcium chloride, with which it forms an adduct).

14.2.39 Formamide

$HCONH_2$	mp 2.55 °C	d_4^{25} 1.129
MW 45.041	bp 210.5 °C(decomposes)	dielectric const. (25 °C) 110.0
	bp 104 °C/10 Torr	

OES 20 ppm A; teratogen.
Completely miscible with water. Readily hydrolysed by acids or bases. Extremely hydroscopic.

Purification
(a) Predry by standing for a few hours at room temperature over anhydrous sodium sulfate (200 g/l), then distil under reduced pressure (water-pump). (b) Crystallise (by freezing-out) in the absence of moisture and carbon dioxide.

A combination of methods (a) and (b) is particularly efficacious.

14.2.40 Glycerine (1,2,3-Trihydroxypropane, Propane-1,2,3-triol)

$HOCH_2CH(OH)CH_2OH$

MW 92.095	mp 18.18 °C	d_4^{25} 1.261
	bp 290.0 °C	dielectric const. (25 °C) 42.5

OES not known.
Not very volatile, probably relatively harmless (cf. 'Ethanol'), unlike Ethylene Glycol (*q.v.*).

No water azeotrope. Completely miscible with water. Extremely hygroscopic and will take up 50% (by weight) of water from the air.

Purification
High-vacuum distillation.

14.2.41 Hexamethylphosphortriamide (HMPA, Hexamethylphosphoramide)

$[(CH_3)_2N]_3PO$

MW 179.204	mp 7.20 °C	d_4^{20} 1.027
	bp 233 °C	dielectric const. (20 °C) 43.30.

Caution: definitely carcinogenic (in animal tests)! May be absorbed through the skin.

Completely miscible with water and many organic solvents; good solvent for salts.

Purification
Predry by standing at room temperature for 24 h over calcium oxide or barium oxide. Then decant and distil under high vacuum over calcium hydride.

Carcinogenic HMPA can usually be replaced as cosolvent for highly reactive nucleophiles or bases by DMPU (*N,N'*-dimethyl-*N,N'*-propyleneurea) [6]. Avoid contact between either HMPA or DMPU and chromium trioxide.

14.2.42 *n*-Hexane

C_6H_{14}
MW 86.178   mp $-94.3\,°C$ d_4^{25} 0.655
              bp $68.7\,°C$ dielectric const. (25 °C) 1.88
OES 50 ppm M
Water azeotrope: 61.6 °C (94.4% hexane). At 25 °C, it is saturated by 0.0111% water; at 25 °C, it is 0.00095% soluble in water.

Commercial hexane in a mixture of hydrocarbons containing a larger or smaller proportion of *n*-hexane together with a variety of other alkanes, and often contaminated by alkenes and less volatile materials.

Purification
Distillation over sodium hydride dispersion. To remove traces of alkenes pretreat by repeatedly shaking with concentrated sulfuric acid (5% by volume), and then washing until neutral.

For UV spectroscopy: Pretreat with sulfuric acid, distil over calcium hydride or sodium hydride and then filter through alumina (basic, activity I, 200 g/l).

14.2.43 Methanol (Methyl Alcohol)

$CH_3 OH$
MW 32.042   mp $-97.6\,°C$ d_4^{25} 0.787
              bp $64.7\,°C$ dielectric const. (25 °C) 32.70
OES 200 ppm A, M. Poisonous if taken orally (5 ml may prove fatal; it should not be substituted for ethanol for oral use!!).

No water azeotrope. Completely miscible with water.

Purification
Place dry magnesium turnings (5 g/l) in a flask and add predried absolute methanol (50 ml). Wait until reaction starts (cloudiness, hydrogen evolution, heat of reaction; a trace of iodine may accelerate this step), then add the methanol requiring purification. Heat slowly to boiling, reflux for 3 h, then distil. Store over 3 Å molecular sieve (10% by weight).

14.2.44 Methyl Acetate

$CH_3CO_2CH_3$
MW 74.080   mp −98.05 °C d_4^{25} 0.928
               bp 57.1 °C   dielectric const. (25 °C) 6.68
OES 10 ppm
At 20 °C, 24% soluble in water, and it dissolves 8% water itself.

Purification
see 'Ethyl Acetate'. Methyl acetate forms an addition compound with calcium chloride, which is thus not a suitable drying agent for it.

14.2.45 4-Methylpentan-2-one (Methyl iso Butyl Ketone)

$(CH_3)_2CHCH_2COCH_3$
MW 100.162   mp −84 °C d_4^{25} 0.802
               bp 116.9 °C   dielectric const. (20 °C) 13.11
OES 50 ppm
Water azeotrope: 87.9 °C (75.7% ketone). At 25 °C it is saturated by 1.9% water.

Purification
Predry over calcium sulfate semihydrate (*ca.* 200 g/l), decant, then fractionally distil.

14.2.46 Monoglyme

see '1,2-Dimethoxyethane'.

14.2.47 Nitrobenzene

$C_6H_5NO_2$
MW 123.112   mp 5.7 °C d_4^{25} 1.198
               bp 210.8 °C   dielectric const. (25 °C) 34.82
OES 1 ppm M. Very readily absorbed through skin.

Water azeotrope: 98.6 °C (12% nitrobenzene). At 20 °C it is saturated by 0.24% water and is 0.2% soluble in water itself.

Purification
Dry over phosphorus pentoxide (*ca.* 50 g/l), decant, then vacuum distil (bp 85 °C/7 Torr).

14.2.48 *n*-Pentane

C_5H_{12} mp −129.7 °C d_4^{25} 0.621
MW 72.151   bp 36.2 °C   dielectric const. (20 °C) 1.84
OES 560 ppm
Water azeotrope: 34.6 °C (98.6% pentane). At 24.8 °C it is saturated by 0. 0120% water.

Purification
see 'Hexane'.

14.2.49 Petroleum Ether (Light Petroleum)

This term is used for hydrocarbon mixtures boiling over a particular temperature range (*e.g.* 40–60 °C, 60–80 °C and 80–100 °C); it is not an ether. It is often possible to use it as a substitute for pentane or hexane (*q.v.*).

Purification
see 'Hexane'.

14.2.50 Propan-l-ol (*n*-Propanol, *n*-Propyl Alcohol)

$CH_3CH_2CH_2OH$
MW 60.097   mp −126.2 °C d_4^{25} 0.804
                bp 97.2 °C   dielectric const. (25 °C) 20.33
OES 200 ppm A
Water azeotrope: 87.65 °C (71.7% alcohol). Completely miscible with water.

Purification
see 'Ethanol'.

14.2.51 Propan-2-ol (*iso*-Propanol, Isopropyl Alcohol)

$(CH_3)_2CHOH$
MW 60.097   mp −88 °C d_4^{25} 0.781
                bp 82.3 °C   dielectric const. (25 °C) 19.92
OES 400 ppm
Water azeotrope: 80.10 °C (88.0% alcohol). Completely miscible with water.

Purification
Predry over calcium sulfate semihydrate (*ca.* 200 g/l), decant, add sodium wire (8 g/l) and heat slowly to boiling. When alkoxide formation is complete, add isopropyl benzoate (35 ml/l), reflux for 3 h, then distil. This provides propan-2-ol suitable for Meerwein-Ponndorf reduction. Other procedures: see 'Ethanol'.

14.2.52 Pyridine

C_5H_5N
MW 79.102   mp −41.5 °C d_4^{25} 0.978
                bp 115.3 °C   dielectric const. (21 °C) 12.4
OES 5 ppm
Water azeotrope: 93.6 °C (58.7% pyridine). Completely miscible with water. Hygroscopic.

Purification
For many purposes it is sufficient to use pyridine that has been stored over solid potassium hydroxide. Traces of potassium hydroxide may be removed by decantation

followed by distillation. Can also be dried over 3 Å molecular sieve or calcium hydride.

14.2.53 Sulfur Dioxide

SO_2

MW 64.06	mp $-75.7\,°C$	d_4^{-10} 1.46
	bp $-10.0\,°C$	dielectric const. $(-20\,°C)$ 17.6

OES 2 ppm

Dry by passing through concentrated sulfuric acid.

14.2.54 Sulfolane

$C_4H_8SO_2$

MW 120.171	mp 28.4 $°C$	d_4^{35} 1.257
	bp 287.3 $°C$ (decomposes)	dielectric const. (30 $°C$) 43.3

T. L. V. Not known. Possibly harmful, so avoid contact with skin.

Purification

Predry at room temperature overnight with potassium hydroxide (500 g/l), decant and distil under high vacuum (use a circulating pump and thermostat at 30 °C to cool the condenser). The pure material is a solid at room temperature.

14.2.55 Tetrahydrofuran (THF)

C_4H_8O

MW 72.108	mp $-108.6\,°C$	d_4^{25} 0.889
	bp 66 $°C$	dielectric const. (25 $°C$) 7.58

OES 200 ppm

Water azeotrope: 63.4 °C (93.3% tetrahydrofuran). Completely miscible with water and most organic solvents.

Purification

Peroxide formation is very rapid so tetrahydrofuran should always be freshly distilled before use. (For removal of peroxides: see 'Diethyl Ether'). Reflux purified tetrahydrofuran with sodium/benzophenone until the blue colour persists, then distil under nitrogen or argon.

14.2.56 Tetrahydronaphthalene (Tetralin)

$C_{10}H_{12}$

MW 132.207	mp $-35.7\,°C$	d_4^{25} 0.966
	bp 207.5 $°C$	dielectric const. (20 $°C$) 2.77

OES not established

Purification

Predry over calcium chloride, reflux for 1 h over sodium (1%), distil and then filter through alumina (basic, activity I, *ca.* 100 g/l). Store under nitrogen; it forms peroxides.

14.2.57 Toluene (Methylbenzene)

$C_6H_5CH_3$

MW 92.142	mp −94.9 °C	d_4^{25} 0.862
	bp 110.6 °C	dielectric const. (25 °C) 2.38

OES 100 ppm M

Water azeotrope: 85 °C (79.8% toluene). At 25 °C it is saturated by 0.0334% water and is itself soluble in water 0.5 g/l at 20 °C.

Purification
see 'Benzene'.

14.2.58 Xylene (Mixture of Isomeric Xylenes)

$C_6H_4(CH_3)_2$

MW 106.169	mp below −45 °C	d_4^{25} *ca.* 0.86
	bp *ca.* 140 °C	dielectric const. (20 °C) *ca.* 2.4

T.L.V. 100 ppm M
Soluble in water 0.1–0.2 g/l at 20 °C.

Purification
Reflux for 1 h over sodium (1%), then distil (see 'benzene').

Bibliography

[1] A useful source of information about molecular sieves is http://www.sigmaaldrich. com/Brands/Aldrich/Tech_Bulletins/AL_143/Molecular_Sieves.html

[2] A short description of the Karl Fischer titration is available: http://www. sigmaaldrich.com/img/assets/4242/fl_analytix3_2000_new.pdf

[3] D. R. Burfield and R. H. Smithers. *J. Org. Chem.*, 1978, 43, 3966.

[4] D. R. Burfield, E. H. Goh, E. H. Ong and R. H. Smithers. *Gazz. Chim. Ital.*, 1983, 113, 841.

[5] D. R. Burfield and R. H. Smithers. *J. Org. Chem.*, 1983, 48, 2420.

[6] T. Mukhopadhyay and D. Seebach, *Helv. Chim. Acta*, 1982, 65, 385.

[8] W. L. F. Armarego and C. Chai. *Purification of Laboratory Chemicals*, Butterworth-Heinemann, Oxford, 2003.

[9] J. A. Ruddick, W. B. Bunger and T. K. Sakano. In A. Weissberger, Ed., *Techniques of Chemistry, Vol 11: Organic Solvents*, 4th edition, John Wiley & Sons, 1986.

Chapter 15

EU *R* and *S* Phrases

15.1 *R* PHRASES: INDICATION OF PARTICULAR RISKS

R1	Explosive when dry
R2	Risk of explosion by shock, friction, fire, or other source of ignition
R3	Extreme risk of explosion by shock, friction, fire, or other source of ignition
R4	Forms very sensitive explosive metallic compounds
R5	Heating may cause explosion
R6	Explosive with or without contact with air
R7	May cause fire
R8	Contact with combustible material may cause fire
R9	Explosive when mixed with combustible material
R10	Flammable
R11	Highly flammable
R12	Extremely flammable
R14	Reacts violently with water
R15	Contact with water liberates extremely flammable gases
R16	Explosive when mixed with oxidising substances
R17	Spontaneously flammable in air
R18	In use may form flammable/explosive vapour-air mixture
R19	May form explosive peroxides
R20	Harmful by inhalation

Practical Organic Synthesis: A Student's Guide   R. Keese, M.P. Brändle and T.P. Toube
© 2006 John Wiley & Sons, Ltd.

R21	Harmful in contact with skin
R22	Harmful if swallowed
R23	Toxic by inhalation
R24	Toxic in contact with skin
R25	Toxic if swallowed
R26	Very toxic by inhalation
R27	Very toxic in contact with skin
R28	Very toxic if swallowed
R29	Contact with water liberates toxic gas
R30	Can become highly flammable in use
R31	Contact with acids liberates toxic gas
R32	Contact with acids liberates very toxic gas
R33	Danger of cumulative effects
R34	Causes burns
R35	Causes severe burns
R36	Irritating to the eyes
R37	Irritating to the respiratory system
R38	Irritating to the skin
R39	Danger of very serious irreversible effects
R40	Possible risk of irreversible effects
R41	Risk of serious damage to the eyes
R42	May cause sensitisation by inhalation
R43	May cause sensitisation by skin contact
R44	Risk of explosion if heated under confinement
R45	May cause cancer
R46	May cause heritable genetic damage
R48	Danger of serious damage to health by prolonged exposure
R49	May cause cancer by inhalation
R50	Very toxic to aquatic organisms
R51	Toxic to aquatic organisms
R52	Harmful to aquatic organisms
R53	May cause long term adverse effects in the aquatic environment
R54	Toxic to flora
R55	Toxic to fauna
R56	Toxic to soil organisms
R57	Toxic to bees
R58	May cause long term adverse effects in the environment
R59	Dangerous to the ozone layer
R60	May impair fertility
R61	May cause harm to the unborn child
R62	Possible risk of impaired fertility
R63	Possible risk of harm to the unborn child
R64	May cause harm to breast-fed babies

R65	Harmful: may cause lung damage if swallowed
R66	Repeated contact may lead to brittle or cracked skin
R67	Vapour can cause drowsiness or numbness

15.2 COMBINATIONS OF *R* NUMBERS

R14/15	Reacts violently with water, liberating extremely flammable gases
R15/29	Contact with water releases toxic, extremely flammable gases
R20/21	Harmful by inhalation and contact with the skin
R20/21/22	Harmful by inhalation, in contact with the skin and if swallowed
R20/22	Harmful by inhalation and if swallowed
R21/22	Harmful in contact with the skin and if swallowed
R23/24	Toxic by inhalation and in contact with the skin
R23/24/25	Toxic by inhalation, in contact with the skin and if swallowed
R23/25	Toxic by inhalation and if swallowed
R24/25	Toxic in contact with the skin and if swallowed
R26/27	Very toxic by inhalation and in contact with the skin
R26/27/28	Very toxic by inhalation, in contact with the skin and if swallowed
R26/28	Very toxic by inhalation and if swallowed
R27/28	Very toxic in contact with the skin and if swallowed
R36/37	Irritating to the eyes and respiratory system
R36/37/38	Irritating to the eyes, respiratory system and skin
R36/38	Irritating to the eyes and skin
R37/38	Irritating to the respiratory system and skin
R39/23/24	Toxic: danger of very serious irreversible effects through inhalation and in contact with the skin
R39/2/24/25	Toxic: danger of very serious irreversible effects through inhalation, in contact with the skin and if swallowed
R39/23/25	Toxic: danger of very serious irreversible effects through inhalation and if swallowed
R39/24	Toxic: danger of very serious irreversible effects in contact with the skin
R39/24/25	Toxic: danger of very serious irreversible effects in contact with the skin and if swallowed
R39/25	Toxic: danger of very serious irreversible effects if swallowed
R39/26	Toxic: danger of very serious irreversible effects through inhalation
R39/26/27	Toxic: danger of very serious irreversible effects through inhalation and in contact with the skin
R39/26/27/28	Very toxic: danger of very serious irreversible effects through inhalation, in contact with the skin and if swallowed

R39/26/28	Very toxic: danger of very serious irreversible effects through inhalation and if swallowed
R39/27	Very toxic: danger of very serious irreversible effects in contact with the skin
R39/27/28	Very toxic: danger of very serious irreversible effects in contact with the skin and if swallowed
R39/28	Very toxic: danger of very serious irreversible effects if swallowed
R40/20/21/22	Harmful: possible risk of irreversible effects through inhalation, in contact with the skin and if swallowed
R40/20	Harmful: possible risk of irreversible effects through inhalation
R40/20/21	Harmful: possible risk of irreversible effects through inhalation and in contact with the skin
R40/20/22	Harmful: possible risk of irreversible effects through inhalation and if swallowed
R40/22	Harmful: possible risk of irreversible effects if swallowed
R40/21	Harmful: possible risk of irreversible effects in contact with the skin
R40/21/22	Harmful: possible risk of irreversible effects through inhalation and in contact with the skin
R42/43	May cause sensitisation by inhalation and skin contact
R48/20/21/22	Harmful: danger of serious damage to health by prolonged exposure through inhalation, in contact with the skin and if swallowed
R48/20	Harmful: danger of serious damage to health by prolonged exposure
R48/20/21	Harmful: danger of serious damage to health by prolonged exposure through inhalation and in contact with the skin
R48/20/22	Harmful: danger of serious damage to health by prolonged exposure through inhalation and if swallowed
R48/21	Harmful: danger of serious damage to health by prolonged exposure in contact with the skin
R48/21/22	Harmful: danger of serious damage to health by prolonged exposure in contact with the skin and if swallowed
R48/22	Harmful: danger of serious damage to health by prolonged exposure if swallowed
R48/23	Toxic: danger of serious damage to health by prolonged exposure through inhalation
R48/23/24/25	Toxic: danger of serious damage to health by prolonged exposure through inhalation, in contact with the skin and if swallowed
R48/23/24	Toxic: danger of serious damage to health by prolonged exposure through inhalation and in contact with the skin
R48/23/25	Toxic: danger of serious damage to health by prolonged exposure through inhalation and if swallowed

R48/24	Toxic: danger of serious damage to health by prolonged exposure in contact with the skin
R48/24/25	Toxic: danger of serious damage to health by prolonged exposure in contact with the skin and if swallowed
R48/25	Toxic: danger of serious damage to health by prolonged exposure if swallowed
R50/53	Very toxic to aquatic organisms, may cause long-term adverse effects in the aquatic environment
R51/53	Toxic to aquatic organisms, may cause long-term adverse effects in the aquatic environment
R52/53	Harmful to aquatic organisms, may cause long-term adverse effects in the aquatic environment

15.3   *S* PHRASES: INDICATION OF SAFETY PRECAUTIONS

S1	Keep locked up
S2	Keep out of reach of children
S3	Keep in a cool place
S4	Keep away from living quarters
S5	Keep contents under…(appropriate liquid specified by manufacturer)
S6	Keep under…(inert gas specified by manufacturer)
S7	Keep container tightly closed
S8	Keep container dry
S9	Keep container in a well ventilated place
S12	Do not keep the container sealed
S13	Keep away from food, drink and animal feeding stuffs
S14	Keep away from…(incompatible materials indicated by manufacturer)
S15	Keep away from heat
S16	Keep away from sources of ignition – No smoking
S17	Keep away from combustible material
S18	Handle and open container with care
S20	When using do no eat or drink
S21	When using do not smoke
S22	Do not breathe dust
S23	Do not breathe gas/fumes/vapour/ spray (specified by manufacturer)
S24	Avoid contact with the skin
S25	Avoid contact with the eyes
S26	In case of contact with the eyes, rinse immediately with plenty of water and seek medical advice
S27	Take off all contaminated clothing immediately

S28	After contact with skin, wash immediately with plenty of...(specified by manufacturer)
S29	Do not empty into drains
S30	Never add water to this product
S33	Take precautionary measures against static discharges
S35	This material and its container must be disposed of in a safe way
S36	Wear suitable protective clothing
S37	Wear suitable gloves
S38	In case of insufficient ventilation, wear suitable breathing apparatus
S39	Wear eye/face protection
S40	To clean floor and all objects contaminated by this material use ... (specified by manufacturer)
S41	In case of fire and/or explosion do not breathe fumes
S42	During fumigation/spraying wear suitable respiratory equipment (specified)
S43	In case of fire, use...(precise equipment to be specified. If water increases the risk add: Never use water)
S45	In case of accident or if you feel unwell, seek medical advice (Show label where possible)
S46	If swallowed seek medical advice immediately and show this container or label
S47	Keep at temperature not exceeding...°C (specified by manufacturer)
S48	Keep wetted with...(specified by manufacturer)
S49	Keep in the original container
S50	Do not mix with (specified by manufacturer)
S51	Use only in well ventilated areas
S52	Not recommended for interior use on large surface areas
S53	Avoid exposure – obtain special instruction before use
S56	Dispose of this material and its container to hazardous or special waste collection point
S57	Use appropriate containment to avoid environmental contamination
S59	Refer to manufacturer/supplier for information on recovery/recycling
S60	This material and/or its container must be disposed of as hazardous waste
S61	Avoid release to the environment. Refer to special instructions/safety data sheet
S62	If swallowed, do not induce vomiting: seek medical advice immediately and show this container or label
S63	If accidentally inhaled: move victim into open air and keep calm
S64	If swallowed wash mouth with water (only if victim is conscious)

15.4 COMBINATIONS OF *S* NUMBERS

S1/2	Keep locked up and out of the reach of children
S3/9/14/49	Keep only in the original container in a cool, well ventilated place away from…(incompatible materials indicated by manufacturer)
S3/9/14	Keep in a cool, well ventilated place away from…(incompatible materials indicated by manufacturer)
S3/9/49	Keep only in the original container in a cool, well ventilated place
S3/14	Keep in a cool, well ventilated place away from…(incompatible materials indicated by manufacturer)
S3/7	Keep container tightly closed in a cool place
S7/8	Keep container tightly closed and dry
S7/9	Keep container tightly closed and in a well ventilated place
S7/47	Keep container tightly closed and at a temperature not exceeding…°C (specified by manufacturer)
S20/21	When using, do not eat, drink or smoke
S24/25	Avoid contact with skin and eyes
S27/28	After contact with skin, take off all contaminated clothing immediately and wash immediately with plenty of…(specified by manufacturer)
S29/35	Do not empty into drains; this material and its container must be disposed of in a safe way
S29/56	Do not empty into drains; dispose of this material and its container to hazardous or special waste collection point
S36/37/39	Wear suitable protective clothing, gloves and eye/face protection
S36/37	Wear suitable protective clothing and gloves
S36/39	Wear suitable protective clothing and eye/face protection
S37/39	Wear suitable gloves and eye/face protection
S47/49	Keep only in the original container at temperature not exceeding…°C (specified by manufacturer)

Index

Practical Organic Synthesis: A Student's Guide   R. Keese, M.P. Brändle and T.P. Toube
© 2006 John Wiley & Sons, Ltd.

With kind thanks to W. F. Farrington for creation of this index.